T0281124

Lecture Notes on Mathematical Olympiad Courses

For Junior Section Vol. 1

Mathematical Olympiad Series

ISSN: 1793-8570

Series Editors: Lee Peng Yee *(Nanyang Technological University, Singapore)*
Xiong Bin *(East China Normal University, China)*

Published

Vol. 1   A First Step to Mathematical Olympiad Problems
by Derek Holton (University of Otago, New Zealand)

Vol. 2   Problems of Number Theory in Mathematical Competitions
by Yu Hong-Bing (Suzhou University, China)
translated by Lin Lei (East China Normal University, China)

Vol. 3   Graph Theory
by Xiong Bin (East China Normal University, China) &
    Zheng Zhongyi (High School Attached to Fudan University, China)
translated by Liu Ruifang, Zhai Mingqing & Lin Yuanqing
    (East China Normal University, China)

Vol. 4   Combinatorial Problems in Mathematical Competitions
by Yao Zhang (Hunan Normal University, P. R. China)

Vol. 5   Selected Problems of the Vietnamese Olympiad (1962–2009)
by Le Hai Chau (Ministry of Education and Training, Vietnam)
    & Le Hai Khoi (Nanyang Technology University, Singapore)

Vol. 6   Lecture Notes on Mathematical Olympiad Courses:
For Junior Section (In 2 Volumes)
by Xu Jiagu

Vol. 7   A Second Step to Mathematical Olympiad Problems
by Derek Holton (University of Otago, New Zealand &
University of Melbourne, Australia)

Vol. 8   Lecture Notes on Mathematical Olympiad Courses:
For Senior Section (In 2 Volumes)
by Xu Jiagu

Xu Jiagu

Former Professor of Mathematics, Fudan University, China

Vol. 8 | Mathematical Olympiad Series

Lecture Notes on Mathematical Olympiad Courses

For Senior Section Vol. 1

 **World Scientific**

Published by

World Scientific Publishing Co. Pte. Ltd.

5 Toh Tuck Link, Singapore 596224

USA office: 27 Warren Street, Suite 401-402, Hackensack, NJ 07601

UK office: 57 Shelton Street, Covent Garden, London WC2H 9HE

British Library Cataloguing-in-Publication Data
A catalogue record for this book is available from the British Library.

Mathematical Olympiad Series — Vol. 8
LECTURE NOTES ON MATHEMATICAL OLYMPIAD COURSES
For Senior Section
(In 2 Volumes)

Copyright © 2012 by World Scientific Publishing Co. Pte. Ltd.

All rights reserved. This book, or parts thereof, may not be reproduced in any form or by any means, electronic or mechanical, including photocopying, recording or any information storage and retrieval system now known or to be invented, without written permission from the Publisher.

For photocopying of material in this volume, please pay a copying fee through the Copyright Clearance Center, Inc., 222 Rosewood Drive, Danvers, MA 01923, USA. In this case permission to photocopy is not required from the publisher.

ISBN-13 978-981-4368-94-0 (pbk)   (Set)
ISBN-10 981-4368-94-6 (pbk)   (Set)

ISBN-13 978-981-4368-95-7 (pbk)   (Vol. 1)
ISBN-10 981-4368-95-4 (pbk)   (Vol. 1)

ISBN-13 978-981-4368-96-4 (pbk)   (Vol. 2)
ISBN-10 981-4368-96-2 (pbk)   (Vol. 2)

Printed in Singapore by World Scientific Printers.

Preface

Although Mathematical Olympiad competitions are carried out by solving problems, the system of Mathematical Olympiads and the related training courses cannot consist only of problem solving techniques. Strictly speaking, it is a system of mathematical advancing education. To guide students, who are interested in and have the potential to enter the world of Olympiad mathematics, so that their mathematical ability can be promoted efficiently and comprehensively, it is important to improve their mathematical thinking and technical ability in solving mathematical problems.

An excellent student should be able to think flexibly and rigorously. Here, the ability to perform formal logic reasoning is an important basic component. However, it is not the main one. Mathematical thinking also includes other key aspects, such as starting from intuition and entering the essence of the subject, through the processes of prediction, induction, imagination, construction and design to conduct their creative activities. In addition, the ability to convert the concrete to the abstract and vice versa is essential.

Technical ability in solving mathematical problems does not only involve producing accurate and skilled-computations and proofs using the standard methods available, but also the more unconventional, creative techniques.

It is clear that the standard syllabus in mathematical education cannot satisfy the above requirements. Hence the Mathematical Olympiad training books must be self-contained basically.

This book is based on the lecture notes used by the editor in the last 15 years for Olympiad training courses in several schools in Singapore, such as Victoria Junior College, Hwa Chong Institution, Nanyang Girls High School and Dunman High School. Its scope and depth significantly exceeds that of the standard syllabus provided in schools, and introduces many concepts and methods from modern mathematics.

The core of each lecture are the concepts, theories and methods of solving mathematical problems. Examples are then used to explain and enrich the lectures, as well as to indicate the applications of these concepts and methods. A number of questions are included at the end of each lecture for the reader to try. Detailed solutions are provided at the end of book.

The examples given are not very complicated so that the readers can understand them easily. However, many of the practice questions at the end of lectures are taken from actual competitions, which students can use to test themselves. These questions are taken from a range of countries, such as China, Russia, the United States of America and Singapore. In particular, there are many questions from China for those who wish to better understand Mathematical Olympiads there. The questions at the end of each lecture are divided into two parts. Those in Part A are for students to practise, while those in Part B test students' ability to apply their knowledge in solving real competition questions.

Each volume can be used for training courses of several weeks with a few hours per week. The test questions are not considered part of the lectures as students can complete them on their own.

Acknowledgments

My thanks to Professor Lee Peng Yee for suggesting the publication of this the book and to Professor Phua Kok Khoo for his strong support. I would also like to thank my friend Mr Fu Ling Chen, lecturer at TJC for his corrections, as well as Zhang Ji and He Yue, the editors of this book at World Scientific Publishing Co. (WSPC). This book would not have been published without their efficient assistance.

Abbreviations and Notations

Abbreviations

AHSME	American High School Mathematics Examination
AIME	American Invitational Mathematics Examination
APMO	Asia Pacific Mathematics Olympiad
ASUMO	Olympics Mathematical Competitions of All the Soviet Union
AUSTRALIA	Australia Mathematical Competitions
AUSTRIA	Austria Mathematical Olympiad
BALKAN	Balkan Mathematical Olympiad
BALTIC WAY	Baltic Way International Mathematical Competition
BELARUS	Belarus Mathematical Olympiad
BMO	British Mathematical Olympiad
BULGARIA	Bulgaria Mathematical Olympiad
CGMO	China Girl's Mathematical Olympiad
CHINA	China Mathematical Competitions for Secondary Schools except for CHNMOL
CHNMO	China Mathematical Olympiad
CHNMOL	China Mathematical Competition for Secondary Schools
CMC	China Mathematical Competition and its preliminary round
CMO	Canada Mathematical Olympiad
CNMO	China Northern Mathematical Olympiad
COLUMBIA	Columbia Mathematical Olympiad
CROATIA	Croatia Mathematical Olympiad
CSMO	China Southeastern Mathematical Olympiad
CWMO	China Western Mathematical Olympiad
CZECH-POLISH-SLOVAK	International Competitions Czech-Polish-Slovak Match
ESTONIA	Estonia Mathematical Olympiad
FINLAND	Finland Mathematical Olympiad
GERMANY	Germany Mathematical Olympiad

GREECE	Greece Mathematical Olympiad
HONG KONG	Hong Kong Mathematical Olympiad
HUNGARY	Hungary Mathematical Competition
IMO	International Mathematical Olympiad
INDIA	India Mathematical Olympiad
IRAN	Iran Mathematical Olympiad
IRE	Ireland Mathematical Olympiad
ITALY	Italy Mathematical Olympiad
JAPAN	Japan Mathematical Olympiad
KOREA	Korea Mathematical Olympiad
KOREAN MC	Korean Mathematical Competition
MACAO	Macao Mathematical Olympiad
MOLDOVA	Moldova Mathematical Olympiad
NEW ZEALAND	New-Zealand Mathematical Olympiad
NORTH-EUROPEAN	North-European Mathematical Olympiad
POLAND	Poland Mathematical Olympiad
ROMANIA	Romania Mathematical Olympiad
RUSMO	All-Russia Olympics Mathematical Competitions
SLOVENIA	Slovenia Mathematical Olympiad
SSSMO	Singapore Secondary Schools Mathematical Olympiads for Senior Section
SMO	Singapore Mathematical Olympiads
SSSMO(J)	Singapore Secondary Schools Mathematical Olympiads for Junior Section
THAILAND	Thailand Mathematical Olympiads
TURKEY	Turkey Mathematical Olympiad
TST	Team Selection Test (including related training tests)
USAMO	United States of American Mathematical Olympiad
VIETNAM	Vietnam Mathematical Olympiad

Notations for Numbers, Sets and Logic Relations

$\mathbb{N}$	the set of positive integers (natural numbers)
$\mathbb{N}_0$	the set of non-negative integers
$\mathbb{Z}$	the set of integers
$\mathbb{Z}^+$	the set of positive integers
$\mathbb{Q}$	the set of rational numbers
$\mathbb{Q}^+$	the set of positive rational numbers
$\mathbb{Q}_0^+$	the set of non-negative rational numbers
$\mathbb{R}$	the set of real numbers
$\mathbb{R}^+$	the set of positive real numbers
$\mathbb{R}_0^+$	the set of non-negative real numbers
$[a, b]$	the closed interval, i.e. all x such that $a \leq x \leq b$
(a, b)	the open interval, i.e. all x such that $a < x < b$
$\Leftrightarrow$	iff, if and only if
$\Rightarrow$	implies
$A \subset B$	A is a subset of B
$A - B$	the set formed by all the elements in A but not in B
$A \cup B$	the union of the sets A and B
$A \cap B$	the intersection of the sets A and B
$a \in A$	the element a belongs to the set A

Contents

Lecture 1

Fractional Equations

Definition 1.1. An equation is called a **fractional equation** if it contains fractional expressions where the unknown variables are appeared in their denominators.

The basic approach for solving a fractional equation is to remove the denominator if possible. However, in many cases this cannot be done by simply multiplying the equation by the least common multiple of the denominators, since it will lead to an equation of a high degree. Therefore the following techniques are often needed.

1. **Use partial fraction techniques**.
2. **Use techniques of telescopic sum**.
3. **Use substitutions of variables or expressions**.
   Sometimes, it is advisable to manipulate the expressions before the substitution is discovered and applied.

Examples

Example 1. Solve equation $\dfrac{x+7}{x+8} - \dfrac{x+8}{x+9} - \dfrac{x+5}{x+6} + \dfrac{x+6}{x+7} = 0.$

Solution The given equation can be simplified to

$$\left(1 - \frac{1}{x+8}\right) - \left(1 - \frac{1}{x+9}\right) - \left(1 - \frac{1}{x+6}\right) + \left(1 - \frac{1}{x+7}\right) = 0,$$

1

so

$$\frac{1}{x+8} - \frac{1}{x+9} = \frac{1}{x+6} - \frac{1}{x+7},$$

$$\frac{1}{x^2+17x+72} = \frac{1}{x^2+13x+42},$$

$$x^2+17x+72 = x^2+13x+42, \quad \therefore x = -\frac{15}{2}.$$

Example 2. Solve the following equation

$$\frac{1}{x^2+2x} + \frac{1}{x^2+6x+8} + \frac{1}{x^2+10x+24} = \frac{1}{5} - \frac{1}{x^2+14x+48}.$$

Solution   By moving the term $-\dfrac{1}{x^2+14x+48}$ to left hand side, it follows that

$$\frac{1}{x^2+2x} + \frac{1}{x^2+6x+8} + \frac{1}{x^2+10x+24} + \frac{1}{x^2+14x+48} = \frac{1}{5},$$

$$\frac{1}{x(x+2)} + \frac{1}{(x+2)(x+4)} + \frac{1}{(x+4)(x+6)} + \frac{1}{(x+6)(x+8)} = \frac{1}{5},$$

$$\frac{1}{2}\left[\left(\frac{1}{x} - \frac{1}{x+2}\right) + \left(\frac{1}{x+2} - \frac{1}{x+4}\right) + \cdots + \left(\frac{1}{x+6} - \frac{1}{x+8}\right)\right] = \frac{1}{5},$$

$$\frac{1}{x} - \frac{1}{x+8} = \frac{2}{5}, \quad \text{so } x^2+8x-20 = 0,$$

then $(x-2)(x+10) = 0$, namely $x_1 = 2$ and $x_2 = -10$.

Example 3. (CHINA/2005) Solve equation $\dfrac{|x-3| - |x+1|}{2|x+1|} = 1.$

Solution   Splitting the left hand side to two terms, then

$$\frac{|x-3|}{2|x+1|} - \frac{1}{2} = 1,$$

$$\left|\frac{x-3}{x+1}\right| = 3,$$

$$\frac{x-3}{x+1} = 3 \quad \text{or} \quad \frac{x-3}{x+1} = -3,$$

$$x-3 = 3x+3 \Rightarrow x = -3 \quad \text{and} \quad x-3 = -3x-3 \Rightarrow x = 0.$$

Thus, $x = -3$ or $x = 0$.

Substitutions of variables or expressions play important role in solving fractional equations, as shown in the following examples.

Example 4. Solve equation $\dfrac{2x^2 + 1}{x + 2} + \dfrac{2x + 4}{2x^2 + 1} = 3.$

Solution   Let $y = \dfrac{2x^2 + 1}{x + 2}$, then the given equation becomes $y + \dfrac{2}{y} = 3.$

$$y + \frac{2}{y} = 3 \Rightarrow y^2 - 3y + 2 = 0 \Rightarrow (y - 2)(y - 1) = 0.$$

When $y = 1$, then $2x^2 - x - 1 = 0$, so $x_1 = 1, x_2 = -\dfrac{1}{2}.$

When $y = 2$, then $2x^2 - 2x - 3 = 0$, so $x_3 = \dfrac{1}{2}(1 - \sqrt{7}), x_4 = \dfrac{1}{2}(1 + \sqrt{7}).$

Example 5. (SSSMO(J)/2006) Suppose that the two roots of the equation

$$\frac{1}{x^2 - 10x - 29} + \frac{1}{x^2 - 10x - 45} - \frac{2}{x^2 - 10x - 69} = 0$$

are α and β. Find the value of $\alpha + \beta$.

Solution   Let $y = x^2 - 10x - 45$, the given equation then becomes

$$\frac{1}{y + 16} + \frac{1}{y} = \frac{2}{y - 24},$$
$$\frac{y + 8}{y^2 + 16y} = \frac{1}{y - 24},$$
$$y^2 - 16y - 192 = y^2 + 16y \Rightarrow y = -6,$$
$$x^2 - 10x - 45 = -6 \Rightarrow x^2 - 10x - 39 = 0.$$

Thus, by Viete's Theorem, $\alpha + \beta = 10.$

Sometimes the manipulations on the given equations are needed for finding the desired substitution, as shown in the following examples.

Example 6. (CHINA/2000) Solve the system for (x, y):

$$\frac{xy}{3x + 2y} = \frac{1}{8}, \qquad \frac{xy}{2x + 3y} = \frac{1}{7}.$$

Solution   By taking reciprocals to two sides of each equation, it follows that

$$\frac{2}{x} + \frac{3}{y} = 8,$$
$$\frac{3}{x} + \frac{2}{y} = 7.$$

Letting $u = \dfrac{1}{x}, v = \dfrac{1}{y}$, the system for (u, v) is

$$2u + 3v = 8,$$
$$3u + 2v = 7.$$

By solving them, it is obtained that $u = 1, v = 2$. Then, returning to (x, y),

$$x = 1 \quad \text{and} \quad y = \frac{1}{2}.$$

Example 7. Solve equation $\dfrac{4x^2 + x + 4}{x^2 + 1} + \dfrac{x^2 + 1}{x^2 + x + 1} = \dfrac{31}{6}$.

Solution   Write the given equation in the form

$$4 + \frac{x}{x^2 + 1} + 1 - \frac{x}{x^2 + x + 1} = 5 + \frac{1}{6},$$

then

$$\frac{1}{x + \dfrac{1}{x}} - \frac{1}{x + \dfrac{1}{x} + 1} = \frac{1}{6}.$$

Let $w = x + \dfrac{1}{x}$, it follows that

$$\frac{1}{w} - \frac{1}{w + 1} = \frac{1}{6},$$
$$w(w + 1) = 6 \Rightarrow w^2 + w - 6 = 0,$$
$$\therefore (w - 2)(w + 3) = 0, \quad \text{namely} \quad w = 2 \text{ or } w = -3.$$

(i) $w = 2 \Rightarrow x + \dfrac{1}{x} = 2$, then $(x - 1)^2 = 0$, so $x_1 = x_2 = 1$.

(ii) $w = -3 \Rightarrow x + \dfrac{1}{x} = -3$, then $x^2 + 3x + 1 = 0$, so

$$x_3 = \frac{-3 - \sqrt{5}}{2}, \quad x_4 = \frac{-3 + \sqrt{5}}{2}.$$

Example 8. Solve equation $\dfrac{x^2}{15} + \dfrac{48}{5x^2} = \dfrac{2x}{3} - \dfrac{8}{x}$.

Solution   When both sides are multiplied by 15, the given equation becomes

$$x^2 + \left(\frac{12}{x}\right)^2 = 10\left(x - \frac{12}{x}\right).$$

Let $y = x - \dfrac{12}{x}$, by completing square it follows that

$$y^2 - 10y + 24 = 0,$$
$$(y - 4)(y - 6) = 0, \quad \therefore \ y = 4 \text{ or } 6.$$

Then

$$y = 4 \Rightarrow x - \frac{12}{x} = 4 \Rightarrow (x - 6)(x + 2) = 0, \text{ i.e. } x_1 = 6, x_2 = -2.$$

$$y = 6 \Rightarrow x - \frac{12}{x} = 6 \Rightarrow x^2 - 6x - 12 = 0, \text{ i.e. } x_3 = 3 - \sqrt{21}, x_4 = 3 + \sqrt{21}.$$

Example 9. Solve equation $2x^2 - 3x + \dfrac{2 - 3x}{x^2} = 1$.

Solution Since $x \neq 0$, the given equation is equivalent to

$$2\left(x^2 + \frac{1}{x^2}\right) - 3\left(x + \frac{1}{x}\right) = 1.$$

Let $y = x + \dfrac{1}{x}$, then $2y^2 - 3y - 4 = 1$, i.e., $(2y - 5)(y + 1) = 0$. Thus, $y = \dfrac{5}{2}$ or -1.

(i) $\quad y = \dfrac{5}{2} \Rightarrow 2x + \dfrac{2}{x} = 5 \Rightarrow 2x^2 - 5x + 2 = 0 \Rightarrow (2x - 1)(x - 2) = 0$

$\Rightarrow x_1 = \dfrac{1}{2}, x_2 = 2.$

(ii) $\quad y = -1 \Rightarrow x + \dfrac{1}{x} = -1 \Rightarrow x^2 + x + 1 = 0$, no real solution.

Example 10. Solve equation $x^2 + \left(\dfrac{x}{x + 1}\right)^2 = 3$.

Solution By completing the square on the left hand side, it follows that

$$\left(x - \frac{x}{x + 1}\right)^2 + \frac{2x^2}{x + 1} = 3,$$
$$\left(\frac{x^2}{x + 1}\right)^2 + 2\frac{x^2}{x + 1} - 3 = 0.$$

Let $y = \dfrac{x^2}{x + 1}$, then $y^2 + 2y - 3 = 0 \Rightarrow (y + 3)(y - 1) = 0$, so $y = -3$ or 1.

(i) $\quad y = -3 \Rightarrow \dfrac{x^2}{x + 1} = -3 \Rightarrow x^2 + 3x + 3 = 0$, no real solution.

(ii) $y = 1 \Rightarrow \dfrac{x^2}{x+1} = 1 \Rightarrow x^2 - x - 1 = 0 \Rightarrow x_1 = \dfrac{1 - \sqrt{5}}{2}, \; x_2 = \dfrac{1 + \sqrt{5}}{2}.$

Testing Questions   (A)

1. Solve the equation $\dfrac{x^3 + 4x^2 + 2x - 8}{x^2 + 2x - 3} = \dfrac{2x^3 + 5x^2 + 4x}{2x^2 + x + 1}.$

2. Solve the equation $\dfrac{x+1}{x+2} + \dfrac{x+6}{x+7} = \dfrac{x+2}{x+3} + \dfrac{x+5}{x+6}.$

3. Solve the equation $\dfrac{1}{(x-1)x} + \dfrac{1}{x(x+1)} + \cdots + \dfrac{1}{(x+9)(x+10)} = \dfrac{11}{12}.$

4. Solve the equation $\dfrac{1}{x^2 + 11x - 8} + \dfrac{1}{x^2 + 2x - 8} + \dfrac{1}{x^2 - 13x - 8} = 0.$

5. Solve the equation $x^2 + \dfrac{x^2}{(x+1)^2} = \dfrac{5}{4}.$

6. Solve the equation $\dfrac{3x^2 + 4x - 1}{3x^2 - 4x - 1} = \dfrac{x^2 + 4x + 1}{x^2 - 4x + 1}.$

7. Find all the real solutions (x, y, z) of the following system

$$\frac{9x^2}{1 + 9x^2} = \frac{3}{2}y, \quad \frac{9y^2}{1 + 9y^2} = \frac{3}{2}z, \quad \frac{9z^2}{1 + 9z^2} = \frac{3}{2}x.$$

8. (RUSMO/1993) Find all positive solutions of the system of fractional equa-

tions

$$x_1 + \frac{1}{x_2} = 4,$$

$$x_2 + \frac{1}{x_3} = 1,$$

$$x_3 + \frac{1}{x_4} = 4,$$

$$x_4 + \frac{1}{x_5} = 1,$$

$$\vdots$$

$$x_{99} + \frac{1}{x_{100}} = 4,$$

$$x_{100} + \frac{1}{x_1} = 1.$$

9. Solve the equation in x: $\dfrac{a+x}{b+x} + \dfrac{b+x}{a+x} = \dfrac{5}{2}$.

10. Solve the equation in x: $\dfrac{a+b}{b+x} + \dfrac{a+c}{x+c} = \dfrac{2(a+b+c)}{x+b+c}$ (where $a+b, a+c, b+c, a+b+c$ are all not zero).

11. Given that the equation $\dfrac{x}{x+1} + \dfrac{x+1}{x} = \dfrac{4x+a}{x(x+1)}$ has only one real root, find the value of real number a.

Testing Questions   (B)

1. Solve equation $\dfrac{x^2 - x + 1}{x^2 + 2} + \dfrac{x+1}{x^2 - x + 1} = 1$.

2. Solve equation $\dfrac{x^3 + 7x^2 + 24x + 30}{x^2 + 5x + 13} = \dfrac{2x^3 + 11x^2 + 36x + 45}{2x^2 + 7x + 20}$.

3. (RUSMO/2005) It is known that there is such a number s such that if real numbers a, b, c, d are all neither 0 nor 1, satisfying $a + b + c + d = s$ and $\dfrac{1}{a} + \dfrac{1}{b} + \dfrac{1}{c} + \dfrac{1}{d} = s$, then $\dfrac{1}{1-a} + \dfrac{1}{1-b} + \dfrac{1}{1-c} + \dfrac{1}{1-d} = s$. Find s.

4. (VIETNAM/2007) Solve the system of equations

$$1 - \frac{12}{y + 3x} = \frac{2}{\sqrt{x}}, \qquad \text{and} \qquad 1 + \frac{12}{y + 3x} = \frac{6}{\sqrt{y}}.$$

5. (BELARUS/2005) Find all triples (x, y, z) with $x, y, z \in (0, 1)$ satisfying

$$\left(x + \frac{1}{2x} - 1\right)\left(y + \frac{1}{2y} - 1\right)\left(z + \frac{1}{2z} - 1\right)$$
$$= \left(1 - \frac{xy}{z}\right)\left(1 - \frac{yz}{x}\right)\left(1 - \frac{zx}{y}\right).$$

6. (NORTH-EUROPEAN/2006) Given that x, y, z are real numbers which are not all equal, satisfying

$$x + \frac{1}{y} = y + \frac{1}{z} = z + \frac{1}{x} = k,$$

   where k is a real number. Find all possible values of k.

7. (GREECE/TST/2009) Find all real solutions (x, y, z) of equation

$$\frac{(x + 2)^2}{y + z - 2} + \frac{(y + 4)^2}{z + x - 4} + \frac{(z + 6)^2}{x + y - 6} = 36,$$

   giving $x, y, z > 3$.

8. (BULGARIA/2004) Given the system of equations

$$\begin{cases} x^2 + y^2 = a^2 + 2, \\ \dfrac{1}{x} + \dfrac{1}{y} = a. \end{cases} \quad (a \in \mathbb{R}).$$

   (a)   Solve the system when $a = 0$;

   (b)   Find range of a such that the system has exactly two solutions.

9. (CZECH-POLISH-SLOVAK/2004) Solve the system of equations

$$\frac{1}{xy} = \frac{x}{z} + 1, \qquad \frac{1}{yz} = \frac{y}{x} + 1, \qquad \frac{1}{zx} = \frac{z}{y} + 1,$$

   where x, y, z are real numbers.

Lecture 2

Higher Degree Polynomial Equations

A polynomial equation is said to be a **higher degree equation** if its degree is greater than 2.

Since the general and systemic approach of polynomial equations involves knowledge of complex numbers, in this chapter, we only discuss the kind of higher degree polynomial equations which can be converted to quadratic equations or can be dealt with special methods, for example by completing squares.

These are usual methods for reducing a higher degree to 1 or 2:

(i) **Factorization Methods**.

Here all kinds of skill for factorization that up to now we have learned are needed, including those by finding roots of a polynomial, by using division of polynomials and by exchanging the positions of variable and parameter, etc.

(ii) **Substitution of variables or expressions**.

Substitution of variables and expressions plays important role in simplifying a polynomial equation and reducing its degree. However it is not always easy to find the right substitution, and, sometimes, manipulation on polynomials (for example, the Binomial expansion) is often necessary for finding the appropriate substitution.

In recent years, some problems on solving *system of higher degree polynomial equations* have appeared in MO competitions of countries. It is possible to reduce the degree of equations by operations on equations and the techniques of inequalities, which are different from those used for solving single higher degree equation.

Example 1. (CROATIA/2005) Find all the real solutions (x, y, z) of the equation

$$4xyz - x^4 - y^4 - z^4 = 1.$$

9

Solution

$$4xyz - x^4 - y^4 - z^4 = 1 \Leftrightarrow (x^4 + 1) + (y^4 + z^4) - 4xyz = 0$$
$$\Leftrightarrow (x^4 - 2x^2 + 1) + (y^4 - 2y^2z^2 + z^4) + 2(x^2 - 2xyz + y^2z^2) = 0$$
$$\Leftrightarrow (x^2 - 1)^2 + (y^2 - z^2)^2 + 2(x - yz)^2 = 0 \Leftrightarrow x = \pm 1, y = \pm z, x = yz$$
$$\Leftrightarrow (x, y, z) = (1, 1, 1); \ (1, -1, -1); \ (-1, 1, -1); \ (-1, -1, 1).$$

Example 2. Find the maximum real roots of equation $3x^7 - x^4 - 30x^5 + 10x^2 + 3x^3 - 1 = 0$.

Solution   By factorizing the left hand side,

$$3x^3(x^4 - 10x^2 + 1) - (x^4 - 10x^2 + 1) = 0,$$
$$(3x^3 - 1)(x^4 - 10x^2 + 1) = 0.$$

(i) $3x^3 - 1 = 0 \Rightarrow x = \dfrac{1}{\sqrt[3]{3}}.$

(ii) $x^4 - 10x^2 + 1 = 0 \Rightarrow x^2 = \dfrac{10 \pm \sqrt{96}}{2} = 5 \pm 2\sqrt{6} = (\sqrt{3} \pm \sqrt{2})^2$, so
$x = \pm(\sqrt{3} \pm \sqrt{2}).$
Thus, the maximum root is $\sqrt{3} + \sqrt{2}$.

Example 3. (CROATIA/2004) Solve equation $(6x + 7)^2(3x + 4)(x + 1) = 6$.

Solution   Multiplying both sides by 12, then

$$(6x + 7)^2[(6x + 8)(6x + 6)] = 72,$$
$$(6x + 7)^2[(6x + 7)^2 - 1] = 72.$$

Let $y = (6x + 7)^2$, then $y \geq 0$ and

$$y^2 - y - 72 = 0 \Rightarrow (y + 8)(y - 9) = 0 \Rightarrow y = 9 \Rightarrow 6x + 7 = \pm 3.$$

(i) $6x + 7 = 3 \Rightarrow x_1 = -\dfrac{2}{3};$

(ii) $6x + 7 = -3 \Rightarrow x_2 = -\dfrac{5}{3}.$

Example 4. Find product of all real roots of the equation

$$x^2 + 2x - \frac{6}{x^2 + 2x - 12} = 13.$$

Solution   Let $y = x^2 + 2x - 12$, then the given equation becomes $y - \dfrac{6}{y} = 1$,
so
$$y^2 - y - 6 = 0 \Rightarrow (y - 3)(y + 2) = 0 \Rightarrow y = 3 \text{ or } y = -2.$$

(i) $y = 3 \Rightarrow x^2 + 2x - 12 = 3 \Rightarrow (x-3)(x+5) = 0 \Rightarrow x_1 = 3, x_2 = -5$;

(ii) $y = -2 \Rightarrow x^2 + 2x - 12 = -2 \Rightarrow x_3 = -1 - \sqrt{11}, x_4 = -1 + \sqrt{11}$.

Thus, product of all the roots is $(3)(-5)(-1-\sqrt{11})(-1+\sqrt{11}) = 150$.

Example 5. Find the value of the maximum real root minus the minimum real root of equation $(x^2 - 5)^4 + (x^2 - 7)^4 = 16$.

Solution Let $y = x^2 - 6$, then $(y+1)^4 + (y-1)^4 = 16$. The binomial expansions

$$(y+1)^4 = y^4 + 4y^3 + 6y^2 + 4y + 1 \text{ and } (y-1)^4 = y^4 - 4y^3 + 6y^2 - 4y + 1$$

yield $y^4 + 6y^2 + 1 = 8$ or $y^4 + 6y^2 - 7 = 0$. Then

$$y^4 + 6y^2 - 7 = 0 \Rightarrow (y^2 + 7)(y^2 - 1) = 0 \Rightarrow y^2 = 1 \Rightarrow y = \pm 1.$$

(i) $y = 1 \Rightarrow x^2 - 6 = 1 \Rightarrow x_1 = -\sqrt{7}, x_2 = \sqrt{7}$;

(ii) $y = -1 \Rightarrow x^2 - 6 = -1 \Rightarrow x_3 = -\sqrt{5}, x_4 = \sqrt{5}$.

Thus, the answer is $\sqrt{7} + \sqrt{7} = 2\sqrt{7}$.

Example 6. Solve equation $3x^4 + 2x^3 - 7x^2 - 2x + 3 = 0$.

Solution It's clear that 0 is not a root, therefore the given equation can be written in the form

$$3\left(x^2 + \frac{1}{x^2}\right) + 2\left(x - \frac{1}{x}\right) - 7 = 0.$$

Let $y = x - \dfrac{1}{x}$ and by completing square on the left hand side, then it follows that $3y^2 + 2y - 1 = 0$. Thus

$$3y^2 + 2y - 1 = 0 \Rightarrow (3y-1)(y+1) = 0 \Rightarrow y = -1 \text{ or } \frac{1}{3}.$$

(i) $y = -1 \Rightarrow x^2 + x - 1 = 0 \Rightarrow x_1 = \dfrac{-1-\sqrt{5}}{2}, \quad x_2 = \dfrac{-1+\sqrt{5}}{2}$;

(ii) $y = \dfrac{1}{3} \Rightarrow 3x^2 - x - 3 = 0 \Rightarrow x_3 = \dfrac{1-\sqrt{37}}{2}, \quad x_4 = \dfrac{1+\sqrt{37}}{2}$.

For factorizing the $f(x, a)$ in the equation $f(x, a) = 0$, sometimes it may be needed to consider the parameter a as a variable and the variable x as a parameter temporarily, as shown in the following example.

Example 7. Solve equation $x^4 - 9x^3 + 2(10-a)x^2 + 9ax + a^2 = 0$ for x, where $a > 0$ is a parameter.

Solution   For the sake of factorizing the left hand side, if considering a as the variable and x as a parameter, it follows that

$$
\begin{aligned}
x^4 &- 9x^3 + 2(10 - a)x^2 + 9ax + a^2 \\
&= a^2 - (2x^2 - 9x)a + (x^4 - 9x^3 + 20x^2) \\
&= a^2 - x(2x - 9)a + x^2(x^2 - 9x + 20) \\
&= a^2 - x(2x - 9)a + x^2(x - 4)(x - 5) \\
&= [a - x(x - 5)][a - x(x - 4)] = (a - x^2 + 5x)(a - x^2 + 4x) \\
&= (x^2 - 5x - a)(x^2 - 4x - a).
\end{aligned}
$$

Therefore the given equation can be written in the form

$$
(x^2 - 5x - a)(x^2 - 4x - a) = 0.
$$

Then

$$
x^2 - 5x - a = 0 \Rightarrow x_1 = \frac{5 - \sqrt{25 + 4a}}{2}, \qquad x_2 = \frac{5 + \sqrt{5 + 4a}}{2},
$$

$$
x^2 - 4x - a = 0 \Rightarrow x_3 = 2 - \sqrt{4 + a}, \qquad x_4 = 2 + \sqrt{4 + a}.
$$

Below are some examples for systems of higher degree equations.

Example 8. (CMO/2003) Find all real positive solutions (if any) to

$$
\begin{aligned}
x^3 + y^3 + z^3 &= x + y + z, \quad \text{and} \\
x^2 + y^2 + z^2 &= xyz.
\end{aligned}
$$

Solution   Without loss of generality, we may assume that $x \geq y \geq z > 0$. From $xyz = x^2 + y^2 + z^2 > 2xy$, we have $z > 2$, so that $x, y, z > 2$ and

$$
(x^3 + y^3 + z^3) - (x + y + z) = x(x^2 - 1) + y(y^2 - 1) + z(z^2 - 1) > 6,
$$

therefore no required solution to the given system.

Example 9. (AUSTRIA/2005) Find all real a, b, c, d, e, f that satisfy the system

$$
\begin{aligned}
4a &= (b + c + d + e)^4, \\
4b &= (c + d + e + f)^4, \\
4c &= (d + e + f + a)^4, \\
4d &= (e + f + a + b)^4, \\
4e &= (f + a + b + c)^4, \\
4f &= (a + b + c + d)^4.
\end{aligned}
$$

Solution   That right hand side of each equation is non-negative implies that a, b, c, d, e, f are all non-negative.

Since these variables are cyclic in the equations, if two are different, there must be two consecutive variables different, say $a < b$ (the discussion for case $a > b$ is similar), then

$$a < b \Rightarrow b + c + d + e < c + d + e + f \Rightarrow b < f \Rightarrow a < f$$
$$\Rightarrow b + c + d + e < a + b + c + d \Rightarrow e < a \Rightarrow e < f$$
$$\Rightarrow f + a + b + c < a + b + c + d \Rightarrow f < d \Rightarrow e < d$$
$$\Rightarrow f + a + b + c < e + f + a + b \Rightarrow c < e \Rightarrow c < d$$
$$\Rightarrow d + e + f + a < e + f + a + b \Rightarrow d < b \Rightarrow d < b < f < d,$$

it leads to a contradiction. Thus, $a = b = c = d = e = f$, and from $4a = (4a)^4$ we have $a = 0$ or $a = \dfrac{1}{4}$. the solution of the system are

$$(a, b, c, d, e, f) = (0, 0, 0, 0, 0, 0) \quad \text{or} \quad (\frac{1}{4}, \frac{1}{4}, \frac{1}{4}, \frac{1}{4}, \frac{1}{4}, \frac{1}{4}).$$

Example 10. (MOLDOVA/TST/2008) Find all solutions $(x, y) \in \mathbb{R} \times \mathbb{R}$ of the following system:

$$x^3 + 3xy^2 = 49, \tag{2.1}$$
$$x^2 + 8xy + y^2 = 8y + 17x. \tag{2.2}$$

Solution   The operation $3x \times (2.2) - (2.1)$ gives

$$2x^3 + 24x^2 y - 51x^2 - 24xy + 49 = 0,$$
$$(x - 1)(2x^2 + 24xy - 49x - 49) = 0.$$

From (2.1), $x = 1 \Rightarrow 3y^2 = 48 \Rightarrow y = \pm 4$, and from (2.1) again, since $x \neq 0$,

$$2x^2 + 24xy - 49x - 49 = 0 \Rightarrow 2x^2 + 24xy - 49x = x^3 + 3xy^2$$
$$\Rightarrow 2x + 24y - 49 = x^2 + 3y^2 \Rightarrow (x - 1)^2 + 3(y - 4)^2 = 0$$
$$\Rightarrow x = 1, y = 4.$$

Thus, the original system has two solutions for (x, y): $(1, 4)$ and $(1, -4)$.

Example 11. (CZECH-POLISH-SLOVAK/2005) Given the positive integer n, find all the non-negative real numbers $x_1, x_2, \ldots, x_n$ such that

$$\begin{cases} x_1 + x_2^2 + x_3^3 + \cdots + x_n^n = n, \\ x_1 + 2x_2 + 3x_3 + \cdots + nx_n = \dfrac{n(n + 1)}{2}. \end{cases}$$

Solution   When the first equation minus the second and then move all the terms to the left hand side, it is obtained that

$$0 = x_1 + x_2^2 + x_3^3 + \cdots + x_n^n - n - \left[x_1 + 2x_2 + 3x_3 + \cdots + nx_n - \frac{n(n+1)}{2} \right]$$

$$= (x_2^2 - 2x_2 + 2 - 1) + (x_3^3 - 3x_3 + 3 - 1) + \cdots + (x_n^n - nx_n + n - 1).$$

For any positive integer $k \geq 2$ and $x \geq 0$, by AM-GM inequality,

$$x^k + k - 1 = x^k + \underbrace{1 + 1 + \cdots + 1}_{k-1} \geq k \sqrt[k]{x^k} = kx,$$

and the equality holds if and only if $x = 1$, so $x_2 = x_3 = \cdots = x_n = 1$. Then the given first equation yields $x_1 = 1$.

Thus, $x_1 = x_2 = \cdots = x_n = 1$.

Testing Questions   (A)

1.   Solve equation $(x + 1)(x + 5)(x + 9) = 231$.

2.   Solve equation $(x + 2)(x + 4)(x + 6)(x + 8) = 48$.

3.   Solve equation $(x - 1)^4 + (x - 7)^4 = 272$.

4.   Solve equation $2x^2 + 7x^3 + 6x^2 + 7x + 2 = 0$.

5.   If the equation $x^4 - (k - 1)x^2 + 2 - k = 0$ has four distinct real roots, find the range of possible values of the real k.

6.   (SWEDEN/2002) Given that real numbers α, β satisfy

$$\alpha^3 - 3\alpha^2 + 5\alpha - 17 = 0 \text{ and}$$
$$\beta^3 - 3\beta^2 + 5\beta + 11 = 0,$$

respectively, find the value of $\alpha + \beta$.

7.   (POLAND/2006) Find all the real solutions of the system

$$\begin{cases} x^5 = 5y^3 - 4z, \\ y^5 = 5z^3 - 4x, \\ z^5 = 5x^3 - 4y. \end{cases}$$

8.   (GERMANY/2003) Find all the pairs (x, y) of two real numbers satisfying the system of equations

$$\begin{aligned} x^3 + y^3 &= 7, \\ xy(x + y) &= -2. \end{aligned}$$

9. (USAMO/TST/2001) Find all pairs of integers (x, y) such that

$$x^3 + y^3 = (x + y)^2.$$

10. (AUSTRALIA/2008) a is a given positive number, and n is an integer $\geq$ 4. Find all the n-tuples $(x_1, x_2, \ldots, x_n)$ of positive numbers satisfying the system

$$\begin{cases} x_1 x_2 (3a - 2x_3) = a^3, \\ x_2 x_3 (3a - 2x_4) = a^3, \\ \cdots \cdots \\ x_{n-2} x_{n-1} (3a - 2x_n) = a^3, \\ x_{n-1} x_n (3a - 2x_1) = a^3, \\ x_n x_1 (3a - 2x_2) = a^3. \end{cases}$$

11. (CZECH-SLOVAKIA-POLAND/2008) In the range of real numbers solve the system of equations

$$\begin{cases} x + y^2 = y^3, \\ y + x^2 = x^3. \end{cases}$$

Testing Questions   (B)

1. (BULGARIA/2003) Find the number of real solutions (x, y, z) to the system

$$\begin{cases} x + y + z &= 3xy, \\ x^2 + y^2 + z^2 &= 3xz, \\ x^3 + y^3 + z^3 &= 3yz. \end{cases}$$

2. (AUSTRIA/2005) Find the conditions on k and d such that the system of equations

$$\begin{cases} x^3 + y^3 = 2 \\ y = kx + d \end{cases}$$

has no real solution (x, y).

3. (GERMANY/2005) Find all real solutions (x, y) of the system of equations

$$x^3 + 1 - xy^2 - y^2 = 0, \tag{2.3}$$
$$y^3 + 1 - x^2 y - x^2 = 0. \tag{2.4}$$

4. (KOREAN MC/2000) Let $\dfrac{3}{4} < a < 1$. Prove that the equation

$$x^3(x + 1) = (x + a)(2x + a)$$

has four distinct real solutions and find these solutions in explicit form.

5. (USAMO/TST/2001) Solve the system of equations:

$$x + \frac{3x - y}{x^2 + y^2} = 3$$
$$y - \frac{x + 3y}{x^2 + y^2} = 0.$$

6. (IRE/2007) Given that r, s, t are all real roots of the polynomial

$$P(x) = x^3 - 2007x + 2002,$$

find the value of $\dfrac{r - 1}{r + 1} + \dfrac{s - 1}{s + 1} + \dfrac{t - 1}{t + 1}$.

7. (CROATIA/TST/2007) In the range of real numbers solve the system

$$\begin{cases} x + y + z = 2, \\ (x + y)(y + z) + (y + z)(z + x) + (z + x)(x + y) = 1, \\ x^2(y + z) + y^2(z + x) + z^2(x + y) = -6. \end{cases}$$

8. (AUSTRALIA/2002) When t is a positive real number, find the number of positive real solutions (a, b, c, d) of the system of equations

$$\begin{cases} a(1 - b^2) = t, \\ b(1 - c^2) = t, \\ c(1 - d^2) = t, \\ d(1 - a^2) = t. \end{cases}$$

9. (USAMO/TST/2001) Find all ordered pairs of real numbers (x, y) for which:

$$(1 + x)(1 + x^2)(1 + x^4) = 1 + y^7,$$
$$\text{and } (1 + y)(1 + y^2)(1 + y^4) = 1 + x^7.$$

10. (CZECH-POLISH-SLOVAK/2008) Find all positive triples (x, y, z) which satisfy the system

$$\begin{cases} 2x^3 = 2y(x^2 + 1) - (z^2 + 1), \\ 2y^4 = 3z(y^1 + 1) - 2(x^2 + 1), \\ 2z^5 = 4x(z^2 + 11) - 3(y^2 + 1). \end{cases}$$

Lecture 3

Irrational Equations

Definition 3.1. An equation is called an **irrational equation** if in the equation some expressions containing unknown variable(s) is (are) under surd form(s).

Methods for Solving irrational Equations

(i)   The key for solving an irrational equation is removing the surd form with unknown variables. In this aspect, taking powers, completing squares, factorization are often applied.

(ii)   Use substitution of variables or expressions is one powerful tool for removing surd forms, and by substitution, the degree of equation can be reduced at the same time.

Examples

Example 1. Solve equation $\sqrt{3x-3} + \sqrt{7x-12} - \sqrt{10x+9} = 0$.

     Solution   Move the third term on the left hand side to the right hand side, then

$$\sqrt{3x-3} + \sqrt{7x-12} = \sqrt{10x+9}.$$

Taking squares to both sides, and simplifying them, then

$$\sqrt{(3x-3)(7x-12)} = 12.$$

Taking squares again to both sides, it follows that

$$21x^2 - 57x - 108 = 0,$$
$$(3x - 12)(7x + 9) = 0,$$
$$\therefore x_1 = 4, \quad x_2 = -\frac{9}{7}.$$

By checking, only $x = 4$ satisfies the original equation.

Example 2. Solve equation $\sqrt[3]{x-1} + \sqrt[3]{x-3} + \sqrt[3]{x-5} = 0$.

Solution   Considering the formula: $a^3 + b^3 + c^3 = 3abc$ if $a + b + c = 0$, the given equation gives

$$(x-1) + (x-3) + (x-5) = 3\sqrt[3]{(x-1)(x-3)(x-5)},$$
$$x - 3 = \sqrt[3]{(x-1)(x-3)(x-5)},$$
$$\sqrt[3]{x-3}[\sqrt[3]{(x-3)^2} - \sqrt[3]{(x-1)(x-5)}] = 0.$$

$\sqrt[3]{x-3} = 0 \Rightarrow x_1 = 3.$
$\sqrt[3]{(x-3)^2} - \sqrt[3]{(x-1)(x-5)} = 0 \Rightarrow 9 = 5$, so no solution.
Thus, $x = 3$ is the unique solution.

Example 3. Solve equation $2\sqrt{x(x+6)} - \sqrt{x} - \sqrt{x+6} = 14 - 2x$.

Solution   Considering $(\sqrt{x} + \sqrt{x+6})^2 = 2x + 2\sqrt{x(x+6)} + 6$, write the given equation in the form

$$(2x + 2\sqrt{x(x+6)} + 6) - (\sqrt{x} + \sqrt{x+6}) - 20 = 0.$$

Let $y = \sqrt{x} + \sqrt{x+6}$, then $y \geq 0$ and $y^2 - y - 20 = 0$.

$$(y-5)(y+4) = 0,$$
$$\therefore y = 5.$$

thus,

$$\sqrt{x} + \sqrt{x+6} = 5 \Rightarrow x + 6 = x + 25 - 10\sqrt{x} \Rightarrow x = \left(\frac{19}{10}\right)^2.$$

Example 4. Solve equation $\sqrt{x+1} + \sqrt{y} + \sqrt{z-4} = \dfrac{x+y+z}{2}$.

Solution   By completing squares,

$$\sqrt{x+1} + \sqrt{y} + \sqrt{z-4} = \frac{x+y+z}{2}$$
$$\Rightarrow (x + 1 - 2\sqrt{x+1} + 1) + (y - 2\sqrt{y} + 1) + (z - 4 - 2\sqrt{z-4} + 1) = 0$$
$$\Rightarrow (\sqrt{x+1} - 1)^2 + (\sqrt{y} - 1)^2 + \left(\sqrt{z-4} - 1\right)^2 = 0,$$

therefore

$$\sqrt{x+1} = 1, \sqrt{y} = 1, \sqrt{z-4} = 1, \text{ i.e., } x = 0, y = 1, z = 5.$$

Example 5. Solve equation $x^2 - 4x - 4 + x\sqrt{x^2 - 2x - 2} = 0$.

Solution Let $y = \sqrt{x^2 - 2x - 2}$, then $2y^2 + xy - x^2 = 0$, and

$$2y^2 + xy - x^2 = 0 \Rightarrow (2y - x)(y + x) = 0, \therefore y = \frac{x}{2} \text{ or } y = -x.$$

(i) $y = \dfrac{x}{2} \Rightarrow \sqrt{x^2 - 2x - 2} = \dfrac{x}{2} \Rightarrow 3x^2 - 8x - 8 = 0$ and $x \geq 0$

$\Rightarrow x_1 = \dfrac{4 + 2\sqrt{10}}{3}$. $\left(x = \dfrac{4 - 2\sqrt{10}}{3} \text{ is not acceptable}\right)$.

(ii) $\sqrt{x^2 - 2x - 2} = -x \Rightarrow -2x - 2 = 0$ and $x \leq 0 \Rightarrow x_2 = -1$.

Thus, the solutions are $x_1 = \dfrac{4 + 2\sqrt{10}}{3}$, $x_2 = -1$.

Example 6. Solve equation $\sqrt{2x + \sqrt{2x - 1}} + \sqrt{2x - \sqrt{2x - 1}} = 2$.

Solution Let $u = \sqrt{2x + \sqrt{2x - 1}}, v = \sqrt{2x - \sqrt{2x - 1}}$, then $u, v \geq 0$ and

$$u + v = 2 \quad \text{and} \quad u - v = \frac{u^2 - v^2}{u + v} = \sqrt{2x - 1},$$

therefore $2u = 2 + \sqrt{2x - 1}$. Returning to x,

$$2\sqrt{2x + \sqrt{2x - 1}} = 2 + \sqrt{2x - 1}$$
$$\Rightarrow 4(2x + \sqrt{2x - 1}) = 4 + 2x - 1 + 4\sqrt{2x - 1}$$
$$\Rightarrow 6x = 3 \Rightarrow x = \frac{1}{2}.$$

It is easy to see that $\dfrac{1}{2}$ satisfies the original equation.

Example 7. Solve the equation $\sqrt{4x^2 + 5x - 2} - \sqrt{4x^2 - 3x - 2} = 2\sqrt{x}$.

Solution It's obvious that $x \neq 0$, hence both sides of the equation can be divided by $\sqrt{x}$ so that the given equation is changed to the form

$$\sqrt{4x - \frac{2}{x} + 5} - \sqrt{4x - \frac{2}{x} - 3} = 2.$$

By the substitution $y = 4x - \dfrac{2}{x}$, it follows that

$$\sqrt{y + 5} - \sqrt{y - 3} = 2.$$

Then

$$\sqrt{y + 5} - \sqrt{y + 3} = 2 \Rightarrow \sqrt{y + 5} = \sqrt{y - 3} + 2$$
$$\Rightarrow y + 5 = y + 1 + 4\sqrt{y - 3} \Rightarrow \sqrt{y - 3} = 1$$
$$\Rightarrow y = 4.$$

Returning to x, then

$$y = 4 \Rightarrow 4x - \frac{2}{x} = 4 \Rightarrow 2x^2 - 2x - 1 = 0 \Rightarrow x = \frac{1 \pm \sqrt{3}}{2}.$$

By checking, the two values are really roots of the original equation.

Example 8. Solve equation $\sqrt[3]{(1 + x)^2} + 15\sqrt[3]{(1 - x)^2} = 8\sqrt[3]{1 - x^2}$.

Solution   It is obvious that $x^2 \neq 1$. Therefore, by moving $\sqrt[3]{1 - x^2}$ to left hand side, the given equation can be written in the form

$$\sqrt[3]{\frac{1 + x}{1 - x}} + 15\sqrt[3]{\frac{1 - x}{1 + x}} = 8.$$

The substitution $y = \sqrt[3]{\dfrac{1 + x}{1 - x}}$ yields $y + \dfrac{15}{y} = 8$ i.e. $y^2 - 8y + 15 = 0$, and its solutions are $y = 3$ or 5.

(i) $y = 3 \Rightarrow \sqrt[3]{\dfrac{1 + x}{1 - x}} = 3 \Rightarrow 1 + x = 27(1 - x) \Rightarrow x_1 = \dfrac{13}{14}.$

(ii) $y = 5 \Rightarrow \sqrt[3]{\dfrac{1 + x}{1 - x}} = 5 \Rightarrow 1 + x = 125(1 - x) \Rightarrow x_2 = \dfrac{62}{63}.$

Example 9. (SSSMO/2001) If x and y are real numbers satisfying

$$\sqrt{x^2 + 2y + 4} + \sqrt{x^2 + x - y + 5} = \sqrt{x^2 + x + 3y + 2} + \sqrt{x^2 + 2x + 3},$$

find the value of $x + y$.

Solution   Since $(x^2 + x + 3y + 2) - (x^2 + 2y + 4) = (x + y - 2) = (x^2 + 2x + 3) - (x^2 + x - y + 5)$, the given equation can be written in the form

$$\sqrt{x^2 + 2y + 4} + \sqrt{x^2 + x - y + 5} = \sqrt{x^2 + 2y + 4 + \delta} + \sqrt{x^2 + x - y + 5 + \delta},$$

where $\delta = x + y - 2$.

For any solution (x, y), the right hand side is a strictly increasing function of δ, so it is greater than the left hand side if $\delta > 0$ and is less than the left hand side if $\delta < 0$. So $\delta = 0$ for any solution (x, y). Thus $x + y = 2$.

The given equation has infinitely many real roots, for example, any pair (x, y) of two positive real numbers with $x + y = 2$ is a solution.

In mathematical competitions, some questions on irrational equations are to analyze the parameters in equations under certain given conditions about their real roots. The Viete's Theorem is often applied for dealing with this kind of problems, as shown in the following examples.

Example 10. (CHINA/2005) Given that the irrational equation in x

$$a\sqrt{x^2} + \frac{1}{2}\sqrt[4]{x^2} - \frac{1}{3} = 0$$

has exactly two distinct real roots. Find the range of the real number a.

Solution   The substitution $w = \sqrt[4]{x^2}$ yields

$$aw^2 + \frac{1}{2}w - \frac{1}{3} = 0. \tag{3.1}$$

The original equation has exactly two distinct real solutions if and only if (3.1) has exactly one positive solution.

When $a = 0$, then $w = \frac{2}{3}$ is the unique positive solution for w.

When $a \neq 0$, by Viete's Theorem, (3.1) has one positive root and one negative root if and only if $a > 0$.

When $a < 0$, then (3.1) may have two equal positive roots, and in that case

$$\Delta = \left(\frac{1}{2}\right)^2 - 4a\left(-\frac{1}{3}\right) = 0,$$

so $a = -\frac{3}{16}$.   Thus, the range of a is $a \geq 0$ or $a = -\frac{3}{16}$.

Example 11. Solve equation $\sqrt{|1 - x|} = kx$, where k is a parameter with $0 < k < 1$.

Solution $kx \geq 0$ gives $x \geq 0$, and the given equation is equivalent to

$$|1 - x| = k^2 x^2.$$

(i)   On the interval $0 \le x \le 1$, the equation becomes $k^2 x^2 + x - 1 = 0$. Its discriminant $\Delta = 1 + 4k^2 > 0$, so it has two roots

$$x_1 = \frac{-1 + \sqrt{1 + 4k^2}}{2k^2}, \quad x_2 = \frac{-1 - \sqrt{1 + 4k^2}}{2k^2}.$$

Here

$$1 < \sqrt{1 + 4k^2} < \sqrt{1 + 4k^2 + 4k^4} = 1 + 2k^2 \Rightarrow 0 < x_1 < 1,$$

but $x_2 < 0$, so only x_1 is a root.

(ii)   On the interval $1 < x$ if $\frac{1}{2} < k$, the equation becomes $k^2 x^2 - x + 1 = 0$ and its discriminant $\Delta = 1 - 4k^2 < 0$, so it has no real roots.

(iii)   On the interval $1 < x$ if $k = \frac{1}{2}$, the equation becomes $x^2 - 4x + 4 = 0$, i.e., $(x - 2)^2 = 0$, so $x_3 = 2$ is a root.

(iv)   On the interval $1 < x$ if $0 < k < \frac{1}{2}$, the discriminant of $k^2 x^2 - x + 1 = 0$ is positive, so

$$x_4 = \frac{1 - \sqrt{1 - 4k^2}}{2k^2}, \quad x_5 = \frac{1 + \sqrt{1 - 4k^2}}{2k^2}.$$

Since $(1 - 2k^2)^2 = 1 - 4k^2 + 4k^4 > 1 - 4k^2$ implies that $1 - 2k^2 > \sqrt{1 - 4k^2}$, so

$$x_4 - 1 = \frac{(1 - 2k^2) - \sqrt{1 - 4k^2}}{2k^2} > 0, \quad \text{i.e. } x_4 > 1,$$

therefore $x_5 > x_4 > 1$. Thus, x_4, x_5 are both roots.

Testing Questions   (A)

1.   (CHNMO/2005) If $\sqrt{7x^2 + 9x + 13} + \sqrt{7x^2 - 5x + 13} = 7x$, find x.

2.   Solve equation
$$x^2 + 4x + 4x\sqrt{x + 3} = 13.$$

3.   (CHINA/2004) Solve equation

$$x - \frac{4}{x} + \frac{3\sqrt{x(x - 2)}}{x} = 2.$$

4. Find the number of real roots of the irrational equation

$$\sqrt{x^2 + 1} + \sqrt{y^2 - 4} + \frac{1}{2}\sqrt{z^2 - 1} = \frac{1}{4}(x^2 + y^2 + z^2 + 5).$$

5. Solve equation

$$\frac{\sqrt{a + x} + \sqrt{a - x}}{\sqrt{a + x} - \sqrt{a - x}} = \frac{a}{x},$$

where $a > 0$ is a parameter.

6. Solve the equation

$$\sqrt{x^2 - 6x + 9} + \sqrt{x^2 + 8x + 16} = 7.$$

7. (CHINA/2005) Solve equation with parameter a

$$\sqrt{x + \sqrt{2x - 1}} + \sqrt{x - \sqrt{2x - 1}} = \sqrt{a},$$

(where $a > 0$), and discuss the solutions.

8. (CROATIA/2008) Find all real solutions of equation

$$x^2 + x + \sqrt{x^2 + x + 7} = 5.$$

9. (CROATIA/TST/2007) Find all real solutions of equation

$$\sqrt{2x + 1} + \sqrt{x + 3} = 3 + \sqrt{x + 7}.$$

10. (VIETNAM/2007) Solve the system of equations

$$\begin{cases} 1 - \dfrac{12}{y + 3x} = \dfrac{2}{\sqrt{x}}, \\ 1 + \dfrac{12}{y + 3x} = \dfrac{6}{\sqrt{y}}. \end{cases}$$

11. (USAMO/TST/2001) Let a, b, and c be real and positive parameters. Solve the equation

$$\sqrt{a + bx} + \sqrt{b + cx} + \sqrt{c + ax} = \sqrt{b - ax} + \sqrt{c - bx} + \sqrt{a - cx}.$$

Testing Questions   (B)

1. Solve equation
    $$\sqrt[3]{x+9} = 2 - \sqrt{x+1}.$$

2. (AUSTRALIA/2004) Find all real solutions of equation
    $$\sqrt{4 - x\sqrt{4 - (x-2)\sqrt{1 + (x-5)(x-7)}}} = \frac{5x - 6 - x^2}{2}.$$

3. (USAMO/TST/2001) Let a and b be given real numbers. Solve the system of equations
    $$\frac{x - y\sqrt{x^2 - y^2}}{\sqrt{1 - x^2 + y^2}} = a, \qquad \frac{y - x\sqrt{x^2 - y^2}}{\sqrt{1 - x^2 + y^2}} = b$$
    for real numbers x and y.

4. (BULGARIA/2007) Solve the system of irrational equations
    $$\begin{cases} \sqrt{x^2 + y^2 - 16(x+y) - 9y + 7} &=& y - 2, \\ x + 13\sqrt[4]{x - y} &=& y + 42. \end{cases}$$

5. (VIETNAM/2009) Solve the system of equations
    $$\begin{cases} \dfrac{1}{\sqrt{1 + 2x^2}} + \dfrac{1}{\sqrt{1 + 2y^2}} &=& \dfrac{2}{\sqrt{1 + 2xy}}, \\ x\sqrt{x(1 - 2x)} + \sqrt{y(1 - 2y)} &=& \dfrac{2}{9}. \end{cases}$$

6. Find all real roots of the irrational equation
    $$\sqrt{3x^2 + x - 1} + \sqrt{x^2 - 2x - 3} = \sqrt{3x^2 + 3x + 5} + \sqrt{x^2 + 3}.$$

7. Find the number of real roots of the irrational equation
    $$|x^2 - 4|x| + 5| = \sqrt{16 - 8x + x^2} - 1.$$

8. (CROATIA/TST/2008) Find all real solutions of equation
    $$(16x^{200} + 1)(y^{200} + 1) = 16(xy)^{100}.$$

Lecture 4

Indicial Functions and Logarithmic Functions

Definition 4.1. A function of the form $y = a^x$ is said to be an **indicial function**, where a is a constant (called *base*) with $a > 0$ and $a \neq 1$, and x is the independent variable with the range $(-\infty, +\infty)$, i.e. $\mathbb{R}$.

$y = a^x$ is a strictly increasing function with the domain $(-\infty, +\infty)$ and the range $(0, +\infty)$ when $a > 1$; and is a strictly decreasing function with the domain $(-\infty, +\infty)$ and the range $(0, +\infty)$ when $0 < a < 1$.

Definition 4.2. The **logarithmic function** $y = \log_a x$, where $a > 0, a \neq 1$ and $x > 0$ is defined as the inverse function of indicial function $y = a^x$. So its domain is $(0, +\infty)$ and its range is $(-\infty, +\infty)$.

Similar to the indicial functions, $y = \log_a x$ is a strictly increasing function when $a > 1$; and is a strictly decreasing function when $0 < a < 1$.

The graph of $y = a^x$ and the graph of $y = \log_a x$ are symmetric with respect to the line $y = x$, as shown in the following diagram.

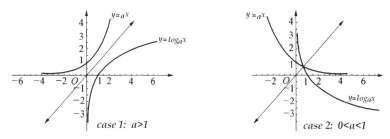

case 1: a>1

case 2: 0<a<1

Basic Operations on Indicial Functions And Logarithmic Functions

For $a > 0$ and $a \neq 1$,

(i) $a^x a^y = a^{x+y}$;
(ii) $(a^x)^y = a^{xy}$;
(iii) $(ab)^x = a^x b^x$.

For $M, N > 0, m, n \in \mathbb{R}, a, b > 0$ and $a \neq 1, b \neq 1$,
(iv)   logarithmic identity: $x = a^{\log_a x}$;
(v) $\log_a MN = \log_a M + \log_a N$;
(vi) $\log_a \dfrac{M}{N} = \log_a M - \log_a N$;
(vii) $\log_a M^n = n \log_a M$;
(viii) $\log_a N = \dfrac{\log_b N}{\log_b a}$. In particular, $\log_a b = \dfrac{1}{\log_b a}$.
(ix) $M^{\log_a N} = N^{\log_a M}$ for $M, N \neq 1$.

Examples

Example 1. (SSSMO/2003) Find the smallest natural number n which satisfies the inequality $12^{200} < n^{300}$.

   Solution   Change the both sides to let them have same power, then compare their bases.

$$12^{200} < n^{300} \Leftrightarrow (144)^{100} < (n^3)^{100} \Leftrightarrow 144 < n^3.$$

Then $5^3 < 144 < 6^3$ implies that $n = 6$.

Example 2. How many integers x satisfy the equation $(x^2 - 2x - 4)^{x^2+3x+2} = 1$?

   Solution   For discussing the equation $f(x)^{g(x)} = 1$, there are three possible cases: (i) $g(x) = 0$ but $f(x) \neq 0$; (ii) $f(x) = 1$; (ii) $f(x) = -1$ and $g(x)$ is an even number.
   (i) When $x^2 + 3x + 2 = 0$ and $x^2 - 2x - 4 \neq 0$, then $x = -1$ or -2;
   (ii) when $x^2 - 2x - 4 = 1$, then $x = 1 - \sqrt{6}$ or $1 + \sqrt{6}$;
   (iii) when $x^2 - 2x - 4 = -1$ and $x^2 + 3x + 2$ is an even number, then $x = 3$ or -1.
   Thus, there are 3 desired values of x.

Example 3. (KOREA MC/2000) Find all real numbers x satisfying the equation

$$2^x + 3^x - 4^x + 6^x - 9^x = 1.$$

   Solution   Setting $2^x = a$ and $3^x = b$, the given equation becomes

$$1 + a^2 + b^2 - a - b - ab = 0.$$

Multiplying both sides of the last equation by 2 and completing squares, then

$$(1-a)^2 + (a-b)^2 + (b-1)^2 = 0.$$

Therefore $a = b = 1$, namely $2^x = 3^x = 1$. So $x = 0$ is the unique solution.

Example 4. (USAMO/TST/2001) Find all real numbers x for which $10^x + 11^x + 12^x = 13^x + 14^x$.

Solution It is easy to check that $x = 2$ is a solution. We claim that it is the only one. In fact, dividing by 13^x on both sides gives

$$\left(\frac{10}{13}\right)^x + \left(\frac{11}{13}\right)^x + \left(\frac{12}{13}\right)^x = 1 + \left(\frac{14}{13}\right)^x.$$

The left hand side is a strictly decreasing function of x and the right hand side is a strictly increasing function of x. Therefore the two curves can have at most one point of intersection.

Example 5. The number of solutions for x in equation

$$5^{2x} - 26 \cdot 5^x + \sqrt{5^{2x} - 26 \cdot 5^x + 26} = -24$$

is

(A) 1   (B) 2   (C) 3   (D) 4   (E) 5.

Solution The substitution $y = \sqrt{5^{2x} - 26 \cdot 5^x + 26}$ yields $y^2 + y - 2 = 0$, therefore $y = 1$. Then

$$\sqrt{5^{2x} - 26 \cdot 5^x + 26} = 1 \Rightarrow (5^x - 1)(5^x - 25) = 0 \Rightarrow x_1 = 0 \quad x_2 = 2.$$

The answer is (B).

Example 6. (SSSMO/2009/Q12) Suppose that a, b and c are real numbers greater than 1. Find the value of

$$\frac{1}{1 + \log_{a^2 b}\left(\frac{c}{a}\right)} + \frac{1}{1 + \log_{b^2 c}\left(\frac{a}{b}\right)} + \frac{1}{1 + \log_{c^2 a}\left(\frac{b}{c}\right)}.$$

Solution The formula for changing base gives $\log_u v = \dfrac{1}{\log_v u}$ if $u, v > 0$ and $u, v \neq 1$, so

$$\frac{1}{1 + \log_{a^2 b}\left(\frac{c}{a}\right)} + \frac{1}{1 + \log_{b^2 c}\left(\frac{a}{b}\right)} + \frac{1}{1 + \log_{c^2 a}\left(\frac{b}{c}\right)}$$

$$= \frac{1}{\log_{a^2 b}(abc)} + \frac{1}{\log_{b^2 c}(abc)} + \frac{1}{\log_{c^2 a}(abc)}$$

$$= \log_{abc}(a^2 b)(b^2 c)(c^2 a) = \log_{abc}(abc)^3 = 3.$$

Example 7. (SLOVANIA/2004) Evaluate 5^a, where $a = \dfrac{\log_7 4(\log_7 5 - \log_7 2)}{\log_7 25(\log_7 8 - \log_7 4)}$.

Solution   First of all, we simplify the index a:

$$a = \frac{\log_7 4(\log_7 5 - \log_7 2)}{\log_7 25(\log_7 8 - \log_7 4)} = \frac{\log_7 2^2 \cdot \log_7 \frac{5}{2}}{\log_7 5^2 \cdot \log_7 2} = \frac{\log_7 \frac{5}{2}}{\log_7 5} = \log_5 \frac{5}{2},$$

therefore $5^a = 5^{\log_5 \frac{5}{2}} = \dfrac{5}{2}$.

Example 8. Solve equation $x + \log_3(3^x - 24) = 4$.

Solution   The given equation yields $\log_3(3^x - 24) = 4 - x$, so $3^x > 24$ and

$$3^x - 24 = 3^{4-x} = \frac{81}{3^x},$$
$$(3^x)^2 - 24 \cdot 3^x - 81 = 0,$$
$$(3^x + 3)(3^x - 27) = 0,$$
$$\because 3^x + 3 > 3, \ \therefore 3^x = 27 = 3^3, \ \text{i.e., } x = 3.$$

Example 9. Solve equation $\log_{30}(x + \sqrt[3]{x}) = \frac{1}{3}\log_3 x$.

Solution   Let $t = \dfrac{1}{3}\log_3 x$, then $x = 3^{3t} = 27^t$ and

$$\log_{30}(27^t + 3^t) = t,$$
$$27^t + 3^t = 30^t,$$
$$\left(\frac{9}{10}\right)^t + \left(\frac{1}{10}\right)^t = 1.$$

Define $f(t) = \left(\dfrac{9}{10}\right)^t + \left(\dfrac{1}{10}\right)^t$, $t \in \mathbb{R}$, then f is a strictly decreasing function on $\mathbb{R}$ and $f(1) = 1$. Thus, $t = 1$, and hence $x = 27^t = 27$.

Example 10. (CROATIA/2008) Solve the system of equations

$$\log_y x + \log_x y = \frac{5}{2}, \tag{4.1}$$
$$x + y = 12. \tag{4.2}$$

Solution   The given equations implies that $x, y > 0$ and $x, y \neq 1$. From (4.1)

$$2(\log_x y)^2 - 5\log_x y + 2 = 0,$$
$$(2\log_x y - 1)(\log_x y - 2) = 0,$$
$$\log_x y = \frac{1}{2} \text{ or } 2 \Rightarrow y = \sqrt{x} \text{ or } x^2.$$

(i) $y = \sqrt{x} \Rightarrow 12 - x = \sqrt{x} \Rightarrow \sqrt{x} = 3 \Rightarrow x = 9, y = 3$;
(ii) $y = x^2 \Rightarrow 12 - x = x^2 \Rightarrow x = 3 \Rightarrow x = 3, y = 9$.
Thus, there are two solutions: $(9, 3)$ and $(3, 9)$.

Testing Questions   (A)

1. (SSSMO/2003) Which of the following numbers is the greatest?

(A) 2^{300} (B) 3^{200} (C) 4^{100} (D) $2^{100} + 3^{100}$ (E) $3^{50} + 4^{50}$.

2. Which is bigger: $\dfrac{15^{2010} + 1}{15^{2011} + 1}$ or $\dfrac{15^{2011} + 1}{15^{2012} + 1}$?

3. (SSSMO/2007/Q11) Suppose that

$$\log_2[\log_3(\log_4 a)] = \log_3[\log_4(\log_2 b)] = \log_4[\log_2(\log_3 c)] = 0.$$

Find the value of $a + b + c$.

4. Given that a, b, c ($a \leq b \leq c$) are natural numbers, and x, y, z, w are real numbers such that

$$a^x = b^{2y} = c^z = 90^w \quad \text{and} \quad \frac{1}{x} + \frac{1}{y} + \frac{1}{z} = \frac{1}{w},$$

prove that $a + b = c$.

5. Solve equations $13^{\log_{11}(x^2 - 10x + 23)} = 7^{\log_{11} 13}$.

6. Evaluate $\log_2(\sqrt{4 + \sqrt{15}} - \sqrt{4 - \sqrt{15}})$.

7. If $x > 0, y > 0$ and $2 \lg(x - 2y) = \lg x + \lg y$, then the value of $x : y$ is

(A) 4;   (B) 1;   (C) 1 or 4;   (D) $\dfrac{1}{4}$;   (E) none of preceding.

8. If $\log_{b^2} x + \log_{x^2} b = 1, b > 0, b \neq 1$ and $x \neq 1$. Then x is equal to

(A) $\dfrac{1}{b^2}$;   (B) $\dfrac{1}{b}$;   (C) b^2;   (D) b;   (E) $\sqrt{b}$.

9. (SSSMO/2010) If $a > b > 1$ and $\dfrac{1}{\log_a b} + \dfrac{1}{\log_b a} = \sqrt{1229}$, find the value of

$$\frac{1}{\log_{ab} b} - \frac{1}{\log_{ab} a}.$$

10. Find the sum of all real numbers x for which
$$(3^x - 9)^3 + (9^x - 81)^3 = (9^x + 3^x - 90)^3.$$

11. (CROATIA/2007) Solve equation $15^{\log_5 x} x^{\log_5 45x} = 1$.

Testing Questions   (B)

1. (USAMO/TST/2001) Find all real numbers x for which
$$\frac{8^x + 27^x}{12^x + 18^x} = \frac{7}{6}.$$

2. (ROMANIA/2003) Given that the positive numbers a, b, c, d satisfy $a > c > d > b > 1$ and $ab > cd$, Prove that the function $f : [0, +\infty) \mapsto \mathbb{R}$ defined by
$$f(x) = a^x + b^x - c^x - d^x$$
is strictly increasing.

3. (CROATIA/2005) Prove that for each positive integer $n \geq 2$,
$$\sum_{k=2}^{n} \left(\log_{\frac{3}{2}} (k^3 + 1) - \log_{\frac{3}{2}} (k^3 - 1) \right) < 1.$$

4. (ROMANIA/1990) Find all real solutions to the equation
$$2^x + 3^x + 6^x = x^2.$$

5. (THAILAND/2003) Find all ordered pairs (x, y) of two real numbers which satisfy the system of equations
$$x^{x+y} = y^{x-y}, \tag{4.3}$$
$$x^2 y = 1. \tag{4.4}$$

6. (SLOVENIA/2005) Calculate the value of
$$\lfloor \log_2 1 \rfloor + \lfloor \log_2 2 \rfloor + \lfloor \log_2 3 \rfloor + \cdots + \lfloor \log_2 256 \rfloor.$$

7. (BULGARIA/2007) Find range of the real parameter a such that the inequality
$$\log_a (a^x + 1) + \frac{1}{\log_{a^{x-1}} a} \leq x - 1 + \log_a (a^2 - 1)$$
holds for all $x \in (0, 1]$.

Lecture 5

Trigonometric Functions

Definition 5.1. Trigonometric Functions For Acute Angles

$$\sin \theta = \frac{a}{c}; \qquad \cos \theta = \frac{b}{c};$$

$$\tan \theta = \frac{a}{b}; \qquad \cot \theta = \frac{b}{a};$$

$$\sec \theta = \frac{c}{b}; \qquad \csc \theta = \frac{c}{a}.$$

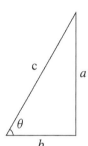

Definition 5.2. Trigonometric Functions For Any Angles

1. $\sin \theta = \dfrac{y}{r}; \quad \cos \theta = \dfrac{x}{r};$

$\tan \theta = \dfrac{y}{x}; \quad \cot \theta = \dfrac{x}{y};$

$\sec \theta = \dfrac{r}{x}; \quad \csc \theta = \dfrac{r}{y}.$

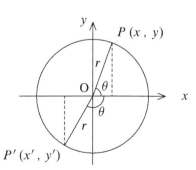

2. Positive and negative angles: Angles measured from x-axis in an anti-clockwise direction are positive angles. Angles measured from x-axis in a clockwise direction are negative angles.

3. The signs of each trigonometric function function in different quadrants are

$$
\begin{array}{c|c}
\begin{aligned} +:\quad & \sin\theta \;\&\; \csc\theta \\[2mm] -:\quad & \text{others} \end{aligned}
& \text{all }(+) \\[4mm]
\hline
\begin{aligned} +:\quad & \tan\theta \;\&\; \cot\theta \\[2mm] -:\quad & \text{others} \end{aligned}
& \begin{aligned} +:\quad & \cos\theta \;\&\; \sec\theta \\[2mm] -:\quad & \text{others} \end{aligned}
\end{array}
$$

4. Each of the six functions are periodic with a common period 2π (or $360°$). Both $\sin x$ and $\cos x$ have range $[-1, 1]$, both $\tan x$ and $\cot x$ has range $(-\infty, +\infty)$, and both $|\sec x|$ and $|\csc x|$ cannot be less than 1.

The functions $\sin x, \tan x, \csc x$ are odd functions, and $\cos x, \cot x, \sec x$ are even functions.

Basic Properties of Trigonometric Functions

1. The graphs of $y = \sin x, y = cos x, y = \tan x, y = \cot x$ are sketched as follows:

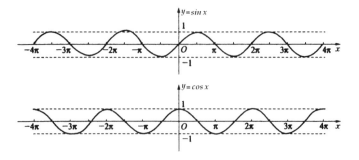

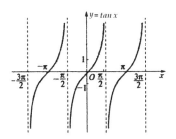
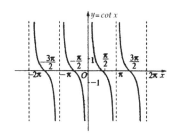

2. Each of the six functions is a periodical function. $y = \sin x$ and $y = \cos x$ take 2π as their minimum period; and each has the range $[-1, 1]$. $y = \tan x$ and $y = \cot x$ take π as their minimum period, and each has the range $(-\infty, +\infty)$.

3. $y = \sin x, x \in \mathbb{R}$ is an odd function, and is an increasing on $[-\frac{\pi}{2}, \frac{\pi}{2}]$, with the range $[-1, 1]$.

   $y = \cos x, x \in \mathbb{R}$ is an even function, and is decreasing on $[0, \pi]$, with the range $[-1, 1]$.

   $y = \tan x$ is an odd function, and is increasing on $(-\frac{\pi}{2}, \frac{\pi}{2})$, with the range $(-\infty, +\infty)$.

   $y = \cot x$ is an odd function, and is decreasing on $(0, \pi)$, with the range $(-\infty, +\infty)$.

4. The functions $y = \sin^{-1} x$ is increasing, it has the domain $[-1, 1]$ and the range $[-\frac{\pi}{2}, \frac{\pi}{2}]$.

   The functions $y = \cos^{-1} x$ is decreasing, it has the domain $[-1, 1]$ and the range $[0, \pi]$.

   The functions $y = \tan^{-1} x$ is increasing, it has the domain $(-\infty, +\infty)$ and the range $(-\frac{\pi}{2}, \frac{\pi}{2})$.

   The functions $y = \cot^{-1} x$ is decreasing, it has the domain $(-\infty, +\infty)$ and the range $(0, \pi)$.

   Below are their sketches, in that order:

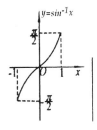
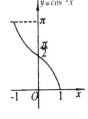
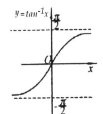
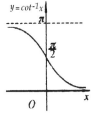

Examples

Example 1. For $\alpha \in \left(0, \frac{\pi}{4}\right)$, arrange $(\sin \alpha)^{\cos \alpha}$, $(\cos \alpha)^{\sin \alpha}$, $(\sin \alpha)^{\sin \alpha}$ in descending order.

Solution Since $0 < \sin \alpha < \cos \alpha < 1$ for $0 < \alpha < \frac{\pi}{4}$,

$$(\cos \alpha)^{\sin \alpha} > (\sin \alpha)^{\sin \alpha} \quad \text{and} \quad (\sin \alpha)^{\sin \alpha} > (\sin \alpha)^{\cos \alpha},$$

therefore

$$(\cos \alpha)^{\sin \alpha} > (\sin \alpha)^{\sin \alpha} > (\sin \alpha)^{\cos \alpha}.$$

Example 2. Compare the values of $\cos(\cos x)$ and $\sin(\sin x)$, if $0 \leq x \leq \pi$.

Solution First of all, $\pi > \frac{\pi}{2} - \cos x > \frac{\pi}{2} - 1 > 0$ and

$$\frac{\pi}{2} - \cos x - \sin x = \frac{\pi}{2} - \sqrt{2} \sin\left(x + \frac{\pi}{4}\right) \geq \frac{\pi}{2} - \sqrt{2} > 0.$$

(i) When $0 \leq x \leq \frac{\pi}{2}$, the angles $\alpha_1 = \frac{\pi}{2} - \cos x, \alpha_2 = \sin x$ are both in $[0, \frac{\pi}{2}]$. Since $\sin u$ is increasing on $[0, \frac{\pi}{2}]$, so

$$\cos(\cos x) - \sin(\sin x) = \sin\left(\frac{\pi}{2} - \cos x\right) - \sin(\sin x) = \sin \alpha_1 - \sin \alpha_2 > 0.$$

(ii) When $\frac{\pi}{2} < x \leq \pi$,

$$\frac{\pi}{2} \leq \alpha_1 = \frac{\pi}{2} - \cos x, \quad 0 \leq \alpha_2 = \sin x \leq 1$$

and

$$\alpha_1 + \alpha_2 = \frac{\pi}{2} - \cos x + \sin x \leq \frac{\pi}{2} + \sqrt{2} < \pi,$$

so $\frac{\pi}{2} \leq \alpha_1 < \pi - \alpha_2$. Hence

$$\cos(\cos x) = \sin(\alpha_1) > \sin(\pi - \alpha_2) = \sin \alpha_2 = \sin(\sin x).$$

Thus, $\cos(\cos x) > \sin(\sin x)$ for $x \in [0, \pi]$.

Example 3. (CMC/2009) In the following functions, the even function with $\frac{\pi}{2}$ as its minimum period is
(A) $y = \sin 2x + \cos 2x$;   (B) $y = \sin 2x \cos 2x$;
(C) $y = \sin^2 x + \cos 2x$;   (D) $y = \sin^2 2x - \cos^2 2x$.

Solution In (A) take $x = \dfrac{\pi}{4}$ then $y = 1$, and take $x = -\dfrac{\pi}{4}$ then $y = -1$, so y is not an even function.

The function in (B) is $y = \dfrac{1}{2}\sin 4x$, so it is not an even function.

The function in (C) is $y = \dfrac{1 + \cos 2x}{2}$, so its minimum period is π.

The function in D) is $y = -\cos 4x$, so it satisfies the requirement.

Example 4. (CMC/2009) Find the range of the function $f(x) = \sin^4 x \cdot \tan x + \cos^4 x \cdot \cot x$.

Solution Since $f(x) = \dfrac{\sin^5 x}{\cos x} + \dfrac{\cos^5 x}{\sin x} = \dfrac{\sin^6 x + \cos^6 x}{\sin x \cos x} = \dfrac{2 - \frac{3}{2}\sin^2 2x}{\sin 2x}$,

if let $t = \sin 2x$, then $t \in [-1, 0) \cup (0, 1]$ and

$$f(x) = g(t) = \frac{2 - \frac{3}{2}t^2}{t} = \frac{2}{t} - \frac{3}{2}t.$$

Since $2/t$ and $-\frac{3t}{2}$ are both decreasing on $[-1, 0)$ and $(0, 1]$, so $g(t) = \dfrac{2}{t} - \dfrac{3}{2}t$ is decreasing on $[-1, 0) \cup (0, 1]$, therefore the range of $g(t)$ is $(-\infty, -\frac{1}{2}] \cup [\frac{1}{2}, +\infty)$.

Thus, the range of $f(x)$ is $\left(-\infty, -\dfrac{1}{2}\right] \cup \left[\dfrac{1}{2}, +\infty\right)$.

Example 5. Prove that the function $f(x) = \cos x^3$, $x \in \mathbb{R}$ is not periodic.

Solution For the sake of contradiction, suppose that T is a period of f. Then

$$f(T) = f(0) = 1 = f(\sqrt[3]{2k\pi}) = f(\sqrt[3]{2k\pi} + T), \quad k \in \mathbb{Z}.$$

Therefore there exists an integer m such that $T^3 = 2m\pi$ or $T = \sqrt[3]{2m\pi}$, where $m \neq 0$. Similarly, for $k = 2m$, there exists $n \in \mathbb{Z}$ such that

$$(\sqrt[3]{4m\pi} + T)^3 = 2n\pi,$$

Thus,

$$(\sqrt[3]{4m\pi} + \sqrt[3]{2m\pi})^3 = 2n\pi \Rightarrow 2m\pi(\sqrt[3]{2} + 1)^3 = 2n\pi$$
$$\Rightarrow m(2 + 3\sqrt[3]{4} + 3\sqrt[3]{2} + 1) = n \Rightarrow \sqrt[3]{4} + \sqrt[3]{2} = \frac{n}{3m} - 1 \Rightarrow \sqrt[3]{4} + \sqrt[3]{2} \in \mathbb{Q}.$$

Let $a = \sqrt[3]{2}$, then $a^2 + a \in \mathbb{Q}$. Since $a^3 = 2 \in \mathbb{Q}$, so $a^3 - 1 = 1 \in \mathbb{Q}$. Then

$$a - 1 = \frac{a^3 - 1}{a^2 + a + 1} \in \mathbb{Q} \Rightarrow a \in \mathbb{Q},$$

which is impossible: It is easy to show that $\sqrt[3]{2} \in \mathbb{Q}^c$ by contradiction. Thus, the contradiction proves the conclusion that T does not exist.

Example 6. For $\theta \in [0, \pi]$ defined $f(\theta) = \sin(\cos \theta)$, $g(\theta) = \cos(\sin \theta)$. If $a = \max\limits_{0 \le \theta \le \pi} f(\theta)$, $b = \min\limits_{0 \le \theta \le \pi} f(\theta)$, $c = \max\limits_{0 \le \theta \le \pi} g(\theta)$, and $d = \min\limits_{0 \le \theta \le \pi} g(\theta)$, then which of the following relations is true?

   (A) $b < d < a < c$;   (B) $d < b < c < a$;

   (C) $b < d < c < a$;   (D) $d < b < a < c$.

Solution   On $[0, \pi]$ the function $y_1 = \cos \theta$ is decreasing, and on $[-1, 1]$ the function $\sin y_1$ is increasing, so $a = \sin 1$, $b = \sin(-1) = -\sin 1$.

On $[0, \pi]$ the function $y_2 = \sin \theta$ has range $[0, 1]$, and the function $\cos y_2$ is decreasing on $[0, 1]$, so $c = \cos 0 = 1$, $d = \cos 1$.

Further, $1 > \dfrac{\pi}{4}$ implies that $a = \sin 1 > \cos 1 = d$.

Thus, $b < d < a < c$, the answer is (A).

Example 7. Given the function $f(x) = \dfrac{6 \sin^4 x - 7 \sin^2 x + 2}{\cos 2x}$. (i) Find the domain and range of f; (ii) determine if f is an odd function or even function.

Solution   (i) $\dfrac{6 \sin^4 x - 7 \sin^2 x + 2}{\cos 2x} = \dfrac{(3 \sin^2 x - 2)(2 \sin^2 x - 1)}{\cos 2x}$

$= 2 - 3 \sin^2 x = 3 \cos^2 x - 1 = \dfrac{3}{2} \cos 2x + \dfrac{1}{2}$.

Since $\cos 2x \ne 0$, so $2x \ne k\pi + \dfrac{\pi}{2}$, i.e., the domain of f is $\left\{ x : x \ne \dfrac{k\pi}{2} + \dfrac{\pi}{4} \right\}$.

The range of f is $\left[-1, \dfrac{1}{2} \right) \cup \left(\dfrac{1}{2}, 2 \right]$.

   (ii)   Since $\cos 2x$ is even function, so is f.

Example 8. (CMC/2008) Given that $f(x) = \cos 2x - 2a(1 + \cos x)$ has minimum value $-\dfrac{1}{2}$, find the value of a.

Solution $f(x) = 2\cos^2 x - 1 - 2a - 2a \cos x = 2 \left(\cos x - \dfrac{a}{2} \right)^2 - \dfrac{1}{2} a^2 - 2a - 1$.

(1) When $a > 2$, the minimum value of f is $1 - 4a < -7$, and the value is obtained when $\cos x = 1$.

(2) When $a < -2$, the minimum value of f is 1, and the value is obtained when $\cos x = -1$.

(3) When $-2 \le a \le 2$, the minimum value of f is $-\frac{1}{2}a^2 - 2a - 1$, and the value is obtained when $\cos x = \frac{a}{2}$.

By solving $-\frac{1}{2}a^2 - 2a - 1 = -\frac{1}{2}$, it is obtained that $a^2 + 4a + 1 = 0$, so

$$a = -2 + \sqrt{3} \quad \text{or} \quad a = -2 - \sqrt{3}.$$

Since $|a| \le 2$, so $a = -2 + \sqrt{3}$.

Testing Questions   (A)

1. (USAMO/TST/2005) Let $0° < \theta < 45°$. Arrange

   $$t_1 = (\tan \theta)^{\tan \theta}, \ t_2 = (\tan \theta)^{\cot \theta}, \ t_3 = (\cot \theta)^{\tan \theta}, \ t_4 = (\cot \theta)^{\cot \theta},$$

   in decreasing order.

2. Prove that $\cos(\sin x) > \sin(\cos x)$ for $x \in [0, \pi]$.

3. Given $f(x) = a \tan x - b \sin x + 2$ and $f(7) = 9$. Find the value of $f(-7)$.

4. For $x \in (0, \frac{\pi}{4})$, let $a = \cos x^{\sin x^{\sin x}}, b = \sin x^{\cos x^{\sin x}}, c = \cos x^{\sin x^{\cos x}}$ and $d = \sin x^{\sin x^{\sin x}}$. Arrange a, b, c, d in ascending order.

5. Find the minimum period of the function $f(x) = \cos(\sin x)$, where $x \in \mathbb{R}$.

6. (CMC/2008) Given that the lengths of three sides BC, CA, AB of $\triangle ABC$ are a, b, c respectively, and a, b, c form a geometric progression. Find the range of the value of the expression

   $$\frac{\sin A \cot C + \cos A}{\sin B \cot C + \cos B}.$$

7. Let $f(x) = \sin^4 x - \sin x \cos x + \cos^4 x, x \in \mathbb{R}$. Find the range of f.

8. (CSMO/2004) Given that for any value of θ in $[0, \frac{\pi}{2}]$, the inequality

   $$\sqrt{2}(2a + 3) \cos \left(\theta - \frac{\pi}{4} \right) + \frac{6}{\sin \theta + \cos \theta} - 2 \sin 2\theta < 3a + 6$$

   always holds. Find the range of a.

9. Prove that the function $f(x) = -x + \sin x$ $(x \in \mathbb{R})$ is not periodic.

10 (CMC/2008) Let $a = \sin(\sin(2008°))$, $b = \sin(\cos 2008°)$, $c = \cos(\sin 2008°)$ and $d = \cos(\cos 2008°)$, then the order of their values is

(A) $a < b < c < d$;   (B) $b < a < d < c$;   (C) $c < d < b < a$;
(D) $d < c < a < b$.

Testing Questions   (B)

1. Prove that when the function $f(x)$ defined on $\mathbb{R}$ is periodic, and its minimal positive period is T_0, then all the period of f must be a multiple of T_0.

2. (RUSMO/2003) Find all angles α for which the three-element set

$$S = \{\sin\alpha, \sin 2\alpha, \sin 3\alpha\}$$

is equal to the set
$$T = \{\cos\alpha, \cos 2\alpha, \cos 3\alpha\}.$$

3. Prove that the function $f(x) = \cos x^4$, $x \in \mathbb{R}$ is not periodic.

4. (CMC/2008) If $2008 = 2^{a_1} + 2^{a_2} + \cdots + 2^{a_n}$, where $\alpha_1, \alpha_2, \ldots, \alpha_n$ are distinct non-negative integers, compare the sizes of the values

$$\sin\sum_{i=1}^{n}\alpha_i, \qquad \cos\sum_{i=1}^{n}\alpha_i, \qquad \tan\sum_{i=1}^{n}\alpha_i.$$

5. (St.Petersburg MC/1999) Find all continuous functions $f : \mathbb{R} \mapsto \mathbb{R}$ such that

$$f(\sin \pi x) = f(x)\cos \pi x \qquad \text{for all } x \in \mathbb{R}.$$

Lecture 6

Law of Sines and Law of Cosines

(I) **The Law of Sines** (or called as **Sine Rule** shortly): In any triangle ABC with the interior angles $\angle A$, $\angle B$ and $\angle C$, it is always true that

$$\frac{a}{\sin A} = \frac{b}{\sin B} = \frac{c}{\sin C},$$

where a, b, c are thee lengths of the sides BC, CA, AB respectively. Further, the **Extended Sine Rule** is always true:

$$a = 2R \sin A, \quad b = 2R \sin B, \quad c = 2R \sin C,$$

where R is the radius of the circumcircle of the $\triangle ABC$.

Proof.
(i) When $\angle A$ is acute, as shown in the diagram (1),
$$a = 2R \sin A' = 2R \sin A.$$
(ii) When $\angle A$ is obtuse, as shown in the diagram (2), then

$$a = 2R \sin A' = 2r \sin(\pi - A) = 2R \sin A. \qquad \square$$

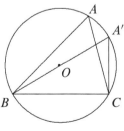

(1)   (2)

39

(II) The Law of Cosines (or called as **Cosine Rule** shortly): In any triangle ABC with the interior angles $\angle A$, $\angle B$ and $\angle C$, it is always true that

$$a^2 = b^2+c^2-2bc \cos A, \quad b^2 = c^2+a^2-2ca \cos B, \quad c^2 = a^2+b^2-2ab \cos C,$$

where a, b, c are thee lengths of the sides BC, CA, AB respectively.

Proof.

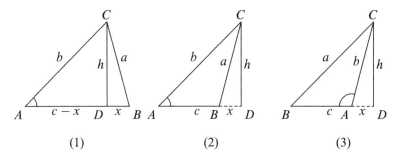

(1)   (2)   (3)

(i) When $\angle A$ is acute, as shown in the diagram (1) or (2), then

$$
\begin{aligned}
a^2 &= h^2 + x^2 = (b \sin A)^2 + (c - b \cos A)^2 \\
&= b^2 \sin^2 A + c^2 + b^2 \cos^2 A - 2bc \cos A \\
&= b^2 + c^2 - 2bc \cos A.
\end{aligned}
$$

(ii) When $\angle A$ is an obtuse angle, as shown in the diagram (3),

$$a^2 = h^2+(c+x)^2 = (b \sin A)^2+(c-b \cos A)^2 = b^2+c^2-2bc \cos A.$$

$\square$

The cosine rule can be expressed in the form

$$\cos A = \frac{b^2 + c^2 - a^2}{2bc}, \quad \cos B = \frac{c^2 + a^2 - b^2}{2ca}, \quad \cos C = \frac{a^2 + b^2 - c^2}{2ab},$$

so the cosine rule means that the interior angles of a given triangle are determined by the lengths of three sides.

The following theorem has wide applications, and can be proven by applying the cosine rule.

Theorem I. (Stewart's Theorem) *For a triangle ABC, if D is an point on the line segment BC such that $BD = p, CD = q$, then*

$$AD^2 = \frac{b^2 p + c^2 q}{p + q} - pq.$$

Proof. As shown in the right figure, applying the cosine rule to $\triangle ABD$ and $\triangle ABC$ gives

$$\frac{c^2 + p^2 - AD^2}{2pc} = \cos B = \frac{a^2 + c^2 - b^2}{2ac}.$$

Note that $a = BC = p + q$, therefore

$$
\begin{aligned}
AD^2 &= c^2 + p^2 + \frac{b^2 p - c^2 p}{a} - pa \\
&= \frac{(p+q)c^2 + b^2 p - pc^2}{a} + p^2 - (p+q)p \\
&= \frac{b^2 p + c^2 q}{p+q} - pq.
\end{aligned}
$$

$\square$

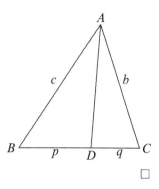

Examples

Example 1. (**Formula for a median**) When AD is the median on the side BC of the $\triangle ABC$, then

$$AD = \frac{1}{2}\sqrt{2b^2 + 2c^2 - a^2}.$$

Solution Applying the Stewart's Theorem to the case $p = q = \frac{1}{2}a$ gives

$$AD^2 = \frac{\frac{1}{2}a}{a}(b^2 + c^2) - \frac{1}{4}a^2 = \frac{1}{4}(2b^2 + 2c^2 - a^2).$$

Therefore $AD = \frac{1}{2}\sqrt{2b^2 + 2c^2 - a^2}$.

Example 2. (**Formula for an angle bisector**) In $\triangle ABC$ when AD is the angle bisector of the $\angle A$, then

$$AD = \frac{2}{b+c}\sqrt{bcs(s-a)},$$

where s is the semi-perimeter of $\triangle ABC$, i.e., $s = \frac{1}{2}(a + b + c)$.

Solution Let $BD = p$, $DC = q$. The angle bisector theorem gives $\frac{p}{q} = \frac{c}{b}$, so

$$p = \frac{ac}{b+c} \quad \text{and} \quad q = \frac{ab}{b+c}.$$

Then the application of the Stewart's Theorem gives

$$
\begin{aligned}
AD^2 &= \frac{b^2ac + c^2ab}{a(b+c)} - \frac{a^2bc}{(b+c)^2} = bc - \frac{a^2bc}{(b+c)^2} = \frac{bc[(b+c)^2 - a^2]}{(b+c)^2} \\
&= \frac{bc(b+c-a)(b+c+a)}{(b+c)^2} = \frac{4bc(s-a)s}{(b+c)^2}.
\end{aligned}
$$

Therefore $AD = \dfrac{2}{b+c}\sqrt{bcs(s-a)}$.

Example 3. (Heron's Formula) The area of $\triangle ABC$, denoted by $[ABC]$, is given by

$$
[ABC] = \sqrt{s(s-a)(s-b)(s-c)},
$$

where $s = \dfrac{1}{2}(a+b+c)$.

Solution   It suffices to find the height h on BC. Let $AD \perp BC$ at D, where D is on the line segment BC, then

$$
c^2 - p^2 = h^2 = b^2 - q^2 \Rightarrow c^2 - b^2 = p^2 - q^2
$$
$$
\Rightarrow p - q = \frac{c^2 - b^2}{a}.
$$

Associate it with $p + q = a$, it follows that

$$
p = \frac{1}{2}\left[\frac{c^2 - b^2}{a} + a\right] = \frac{c^2 + a^2 - b^2}{2a},
$$
$$
q = \frac{1}{2}\left[a - \frac{c^2 - b^2}{a}\right] = \frac{a^2 + b^2 - c^2}{2a},
$$

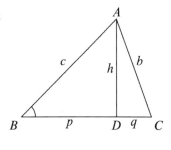

then applying the Stewart's Theorem gives

$$
\begin{aligned}
h^2 &= c^2 - p^2 = \frac{4a^2c^2 - (c^2 + a^2 - b^2)^2}{4a^2} \\
&= \frac{(2ac - c^2 - a^2 + b^2)(2ac + c^2 + a^2 - b^2)}{4a^2} \\
&= \frac{1}{4a^2}\left[b^2 - (c-a)^2\right]\cdot\left[(c+a)^2 - b^2\right] \\
&= \frac{1}{4a^2}(b - a + c)(b + a - c)(a + c - b)(a + c + b) \\
&= \frac{4(s-a)(s-c)(s-b)s}{a^2}.
\end{aligned}
$$

Therefore

$$
h = \frac{2}{a}\sqrt{s(s-a)(s-b)(s-c)} \quad \text{and} \quad [ABC] = \sqrt{s(s-a)(s-b)(s-c)}.
$$

Note: In above proof of Heron's Formula, it was assumed that D is on the line segment BC, but it is easy to see that the formula is still true for any obtuse triangles.

Example 4. (BELARUS/2003) Given $\dfrac{\sin a}{\sin b} = \dfrac{\sin c}{\sin d} = \dfrac{\sin(a-c)}{\sin(b-d)}$, where $a, b, c,$ $d \in (0, \pi)$. Prove that $a = b, c = d$.

Solution Since $a = c$ implies that $a = c = 0$, but it is impossible, so $a \neq c$. Suppose that $a > c$, then the given equalities implies that $b > d$, and

$$\frac{\sin(\pi - a)}{\sin(\pi - b)} = \frac{\sin c}{\sin d} = \frac{\sin(a - c)}{\sin(b - d)}.$$

Let the $\triangle ABC$ satisfy $\angle A = \pi - a, \angle B = c, \angle C = a - c$, and $\triangle DEF$ satisfy $\angle D = \pi - b, \angle E = d, \angle F = b - d$. Since $\dfrac{\sin A}{\sin D} = \dfrac{\sin B}{\sin E} = \dfrac{\sin C}{\sin F}$, by sine rule it follows that

$$\frac{BC}{EF} = \frac{AC}{DF} = \frac{AB}{DE},$$

therefore $\triangle ABC \sim \triangle DEF$, $\angle A = \angle D, \angle B = \angle E$, i.e. $a = b, c = d$.

Example 5. If the lengths a, b, c of three sides of $\triangle ABC$ satisfy $2b = a + c$, find the value of $5 \cos A - 4 \cos A \cos C + 5 \cos C$.

Solution By the sine rule, the relation $2b = a + c$ implies that $2 \sin B = \sin A + \sin C$. Then

$$\sin(A + C) = \frac{\sin A + \sin C}{2}$$
$$\Rightarrow 2 \sin \frac{A + C}{2} \cos \frac{A + C}{2} = \sin \frac{A + C}{2} \cos \frac{A - C}{2}$$
$$\Rightarrow 2 \cos \frac{A + C}{2} = \cos \frac{A - C}{2}.$$

Therefore

$$5 \cos A - 4 \cos A \cos C + 5 \cos C = 5(\cos A + \cos C) - 4 \cos A \cos C$$
$$= 10 \cos \frac{A + C}{2} \cos \frac{A - C}{2} - 2[\cos(A + C) + \cos(A - C)]$$
$$= 10 \cos \frac{A + C}{2} \cos \frac{A - C}{2} - 4[\cos^2 \frac{A + C}{2} + \cos^2 \frac{A - C}{2} - 1]$$
$$= 20 \cos^2 \frac{A + C}{2} - 4[\cos^2 \frac{A + C}{2} + 4 \cos^2 \frac{A + C}{2} - 1] = 4.$$

Note: In the proof, applying the sine rule to convert the relation among sides to that among angles plays important role.

Example 6. (CROATIA/2004) Prove that the inequality

$$\frac{\cos A}{a^3} + \frac{\cos B}{b^3} + \frac{\cos C}{c^3} \geq \frac{3}{2abc}$$

holds for any triangle ABC, where a, b, c are the lengths of three sides, and $\angle A, \angle B, \angle C$ are their opposite inner angles respectively.

Solution   The cosine rule and the inequality $x + \dfrac{1}{x} \geq 2$ for $x > 0$ give that

$$\frac{\cos A}{a^3} + \frac{\cos B}{b^3} + \frac{\cos C}{c^3} = \frac{b^2 + c^2 - a^2}{2a^3bc} + \frac{a^2 + c^2 - b^2}{2b^3ca} + \frac{a^2 + b^2 - c^2}{2c^3ab}$$

$$= \frac{1}{2abc} \left\{ \left[\left(\frac{a}{b}\right)^2 + \left(\frac{b}{a}\right)^2 \right] + \left[\left(\frac{b}{c}\right)^2 + \left(\frac{c}{b}\right)^2 \right] + \left[\left(\frac{c}{a}\right)^2 + \left(\frac{a}{c}\right)^2 \right] - 3 \right\}$$

$$\geq \frac{1}{2abc}(2 + 2 + 2 - 3) = \frac{3}{2abc}.$$

Example 7. (CMC/2008) Let a, b, c be lengths of three sides of $\triangle ABC$, and $b^2 = ac$. If $\angle B = x$ and $f(x) = \sin\left(4x - \dfrac{x}{6}\right) - \dfrac{1}{2}$, find the range of $f(x)$.

Solution   The cosine rule gives

$$\cos x = \frac{a^2 + c^2 - b^2}{2ac} \geq \frac{2ac - ac}{2ac} = \frac{1}{2}.$$

Since $0 < x < \pi$, so $0 < x \leq \dfrac{\pi}{3}$ and $-\dfrac{\pi}{6} < 4x - \dfrac{\pi}{6} \leq \dfrac{7\pi}{6}$. Therefore $-\dfrac{1}{2} \leq \sin\left(4x - \dfrac{x}{6}\right) \leq 1$ and the range of $f(x)$ is $\left[-1, \dfrac{1}{2}\right]$.

Example 8. Given that in the $\triangle ABC$, $a\cos A = b\cos B$. Determine the shape of the $\triangle ABC$.

Solution 1   By the cosine rule,

$$a\cos A = b\cos B \Leftrightarrow a \cdot \frac{b^2 + c^2 - a^2}{2bc} = b \cdot \frac{a^2 + c^2 - b^2}{2ac}$$

$$\Leftrightarrow a^2(b^2 + c^2 - a^2) = b^2(a^2 + c^2 - b^2) \Leftrightarrow a^2c^2 - a^4 - b^2c^2 + b^4 = 0$$

$$\Leftrightarrow (a^2 - b^2)(c^2 - a^2 - b^2) = 0 \Leftrightarrow a = b \text{ or } a^2 + b^2 = c^2.$$

Thus, $\triangle ABC$ is isosceles or right-angled triangle.

Solution 2   The sine rule gives $2R\sin A\cos A = 2R\sin B\cos B$, so $\sin 2A = \sin 2B$.

Since $0 < 2A, 2B < 2\pi$, so $A = B$ or $2A = \pi - 2B$. $A = B$ implies that $\triangle ABC$ is an isosceles triangle.

If $2A = \pi - 2B$, then $A + B = \dfrac{\pi}{2}$, so $C = \dfrac{\pi}{2}$, so $\triangle ABC$ is a right triangle.

Thus, $\triangle ABC$ is isosceles or right-angled triangle.

Note: In the proofs, we focus on the relation of sides by using the cosine rule; and we focus on the relation of angles by using the sine rule.

Testing Questions   (A)

1. In $\triangle ABC$, $AB = AC$, the angle bisector BD of the $\angle B$ intersects AC at D, and $BC = BD + AD$. Find $\angle A$ in degrees.

2. In $\triangle ABC$, $AB = 4, BC = 7, AC = 10$, D is on AC such that $BD = 4$. Find $AD : DC$.

3. (CROATIA/2007) For the $\triangle ABC$ with semi-perimeter p, prove that
$$p^2 = b^2 \cos^2 \frac{C}{2} + c^2 \cos^2 \frac{B}{2} + 2bc \cos \frac{B}{2} \cdot \cos \frac{C}{2} \cdot \cos \frac{B+C}{2}.$$

4. In $\triangle ABC$, $\angle A = 45°, b = 2$ and $[ABC] = 2$. Find the value of
$$\frac{a+b+c}{\sin A + \sin B + \sin C}.$$

5. Given that the circumcenter of acute triangle ABC is D, the circle passing through A, B, D intersects AC and BC at M and N respectively. Prove that the radius of the circumcircle of $\triangle MNC$ is equal to the radius of the circumcircle of $\triangle ABD$.

6. (CROATIA/2008) In the $\triangle ABC$, the angle bisector BK of $\angle B$ intersect AC at K. If $BC = 2, CK = 1, BK = \dfrac{3}{\sqrt{2}}$, find area of $\triangle ABC$.

7. (CMC/2008) Let a, b, c be the lengths of sides BC, CA, AB of the acute triangle ABC, and $\sin A = 2a \sin B$. if the circumradius of the $\triangle ABC$ is $\dfrac{\sqrt{3}}{6}$, find the range of the perimeter of the $\triangle ABC$.

8. (USAMO/TST/2005) In triangle ABC, show that
   (a) $4R = \dfrac{abc}{[ABC]}$;   (b) $2R^2 \sin A \sin B \sin C = [ABC]$;
   (c) $2R \sin A \sin B \sin C = r(\sin A + \sin B + \sin C)$.

Testing Questions   (B)

1. (CROATIA/2004) If the lengths of three sides of a triangle a, b, c satisfy

$$\frac{b}{a} = \frac{|b^2 + c^2 - a^2|}{bc}, \quad \frac{c}{b} = \frac{|c^2 + a^2 - b^2|}{ca}, \quad \frac{a}{c} = \frac{|a^2 + b^2 - c^2|}{ab},$$

    find the three angles.

2. (SMO/2009) Let ABC be a triangle with sides $AB = 7$, $BC = 8$ and $AC = 9$. A unique circle can be draw touching the side AC and the lines BA produced and BC produced. Let D be the centre of this circle. Find the value of BD^2.

3. (CMC/2009) One diagonal AC of the cyclic quadrilateral $ABCD$ partitions the interior angles A and C as four angles $\alpha_1, \alpha_2, \alpha_3, \alpha_4$, as shown in the given graph. Prove that

$$\sin(\alpha_1 + \alpha_2)\sin(\alpha_2 + \alpha_3)\sin(\alpha_3 + \alpha_4)\sin(\alpha_4 + \alpha_1)$$
$$\geq 4\sin\alpha_1\sin\alpha_2\sin\alpha_3\sin\alpha_4.$$

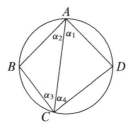

4. (SERBIA/2009) Given that the triangle ABC has unequal sides. The angle bisectors of $\angle BAC$ and $\angle ABC$ intersect their opposite sides at the points D and E respectively. Let $\angle BAC = \alpha$, $\angle ABC = \beta$. Prove that the angle included by the lines DE and AB is not greater than $\dfrac{|\alpha - \beta|}{3}$.

5. (USAMO/TST/2005) In triangle ABC, show that

    (a) $r = 4R \sin \dfrac{A}{2} \sin \dfrac{B}{2} \sin \dfrac{C}{2}$;

    (b) $a \cos A + b \cos B + c \cos C = \dfrac{abc}{2R^2}$.

Lecture 7

Manipulations of Trigonometric Expressions

Trigonometric formulae have widespread applications in trigonometry and geometry, in particular, it is a powerful tool for solving geometric complicated calculate problems. In this lecture we are not going to discuss these complicated problems, but focus on the following aspects only.

1. The evaluation and simplification of trigonometric expressions;
2. The evaluation of trigonometric series;
3. Trigonometric identities of triangles.

Examples

Example 1. (CMC/2009) Evaluate $\cos 10° \cos 50° \cos 70° + \sin 10° \sin 50° \sin 70°$.

Solution Rewrite the given $\cos 10° \cos 50° \cos 70° + \sin 10° \sin 50° \sin 70°$ to

$$\sin 20° \sin 40° \sin 80° + \cos 20° \cos 40° \cos 80°,$$

then

$$8 \sin 20° \sin 40° \sin 80° = 4(\cos 20° - \cos 60°) \sin 80°$$
$$= 4 \sin 80° \cos 20° - 2 \sin 80° = 2(\sin 100° + \sin 60°) - 2 \sin 80°$$
$$= 2 \sin 60° = \sqrt{3},$$

$$8 \cos 20° \cos 40° \cos 80° = \frac{1}{\sin 20°} \cdot 8 \sin 20° \cos 20° \cos 40° \cos 80°$$

$$= \frac{1}{\sin 20°} \cdot 4 \sin 40° \cos 40° \cos 80° = \frac{1}{\sin 20°} \cdot 2 \sin 80° \cos 80°$$

$$= \frac{1}{\sin 20°} \cdot \sin 160° = 1,$$

therefore $\cos 10° \cos 50° \cos 70° + \sin 10° \sin 50° \sin 70° = \dfrac{\sqrt{3} + 1}{8}$.

Example 2. (SLOVENIA/2004) Evaluate $\sin^8 75° - \cos^8 75°$.

Solution   Applying the factorization of $a^8 - b^8$ gives

$$
\begin{aligned}
\sin^8 75° - \cos^8 75° &= (\sin^4 75° - \cos^4 75°)(\sin^4 75° + \cos^4 75°) \\
&= (\sin^2 75° - \cos^2 75°)(\sin^2 75° + \cos^2 75°) \\
&\quad \cdot[(\sin^2 75° + \cos^2 75°)^2 - 2\sin^2 75° \cdot \cos^2 75°] \\
&= -\cos 150° \left(1 - \frac{1}{2}\sin^2 150°\right) \\
&= \frac{\sqrt{3}}{2}\left(1 - \frac{1}{8}\right) = \frac{7\sqrt{3}}{16}.
\end{aligned}
$$

Example 3. (CMC/2008) Given $\dfrac{\sin(\alpha + \beta)}{\sin(\alpha - \beta)} = 3$, find the value of $\dfrac{\tan \alpha}{\tan \beta}$.

Solution   The given equality gives

$$\sin \alpha \cos \beta + \cos \alpha \sin \beta = 3(\sin \alpha \cos \beta - \cos \alpha \sin \beta),$$

namely

$$\sin \alpha \cos \beta = 2 \cos \alpha \sin \beta, \quad \therefore \ \frac{\tan \alpha}{\tan \beta} = \frac{\sin \alpha \cos \beta}{\cos \alpha \sin \beta} = 2.$$

Example 4. (SMO/2010) If $\cot \alpha + \cot \beta + \cot \gamma = -\dfrac{4}{5}$, $\tan \alpha + \tan \beta + \tan \gamma = \dfrac{17}{6}$ and $\cot \alpha \cot \beta + \cot \beta \cot \gamma + \cot \gamma \cot \alpha = -\dfrac{17}{5}$, find the value of $\tan(\alpha + \beta + \gamma)$.

Solution   Let $x = \tan \alpha$, $y = \tan \beta$ and $z = \tan \gamma$. Then

$$\frac{xy + yz + zx}{xyz} = -\frac{4}{5}, \tag{7.1}$$

$$x + y + z = \frac{17}{6}, \tag{7.2}$$

$$\frac{x + y + z}{xyz} = -\frac{17}{5}. \tag{7.3}$$

(7.2) $\div$ (7.3) gives $xyz = -\dfrac{5}{6}$, then (7.1) yields $xy + yz + zx = \dfrac{2}{3}$. Thus

$$
\begin{aligned}
\tan(\alpha + \beta + \gamma) &= \frac{\tan(\alpha + \beta) + \tan \gamma}{1 - \tan(\alpha + \beta)\tan \gamma} \\
&= \frac{(\tan \alpha + \tan \beta + \tan \gamma - \tan \alpha \tan \beta \tan \gamma)/(1 - \tan \alpha \tan \beta)}{(1 - \tan \alpha \tan \beta - \tan \alpha \tan \gamma - \tan \beta \tan \gamma)/(1 - \tan \alpha \tan \beta)} \\
&= \frac{x + y + z - xyz}{1 - (xy + yz + zx)} = \frac{(17/6) + (5/6)}{1 - (2/3)} = 11.
\end{aligned}
$$

Example 5. Prove that $\displaystyle\sum_{k=1}^{n} \sin(\alpha + k\beta) = \dfrac{\sin \frac{n\beta}{2} \sin(\alpha + \frac{n+1}{2}\beta)}{\sin \frac{1}{2}\beta}$.

Solution Let $S = \sin(\alpha + \beta) + \sin(\alpha + 2\beta) + \cdots + \sin(\alpha + n\beta)$. Then

$$S \cdot \sin \frac{\beta}{2} = \sum_{r=1}^{n} \sin \frac{\beta}{2} \sin(\alpha + r\beta)$$

$$= \frac{1}{2} \sum_{r=1}^{n} \left[\cos\left(\alpha + (r - \frac{1}{2})\beta\right) - \cos\left(\alpha + (r + \frac{1}{2})\beta\right)\right]$$

$$= \frac{1}{2} \left[\cos\left(\alpha + \frac{1}{2}\beta\right) - \cos\left(\alpha + (n + \frac{1}{2})\beta\right)\right]$$

$$= \sin \frac{n\beta}{2} \sin\left(\alpha + \frac{n+1}{2}\beta\right), \quad \therefore S = \frac{\sin \frac{n\beta}{2} \sin\left(\alpha + \frac{n+1}{2}\beta\right)}{\sin \frac{1}{2}\beta}.$$

Example 6. Prove that for any natural number n and real number x with $2^k x \neq (m + \frac{1}{2})\pi$ for all $k \in \mathbb{N}$ and $m \in \mathbb{Z}$,

$$\tan x + 2 \tan 2x + 2^2 \tan 2^2 x + \cdots + 2^n \tan 2^n x = \cot x - 2^{n+1} \cot 2^{n+1} x.$$

Solution For any real $\alpha \neq (m + \frac{1}{2})\pi$,

$$\tan 2\alpha = \frac{2 \tan \alpha}{1 - \tan^2 \alpha} \Rightarrow \cot 2\alpha = \frac{1}{2}(\cot \alpha - \tan \alpha)$$

$$\Rightarrow \tan \alpha = \cot \alpha - 2 \cot 2\alpha \Rightarrow 2^k \tan 2^k x = 2^k \cot 2^k x - 2^{k+1} \cot 2^{k+1} x$$

for $k = 0, 1, 2, \cdots, n$. Thus,

$$\tan x + 2 \tan 2x + 2^2 \tan 2^2 x + \cdots + 2^n \tan 2^n x$$
$$= (\cot x - 2 \cot 2x) + (2 \cot 2x - 2^2 \cot 2^2 x) + \cdots + (2^n \cot 2^n x - 2^{n+1} \cot 2^{n+1} x)$$
$$= \cot x - 2^{n+1} \cot 2^{n+1} x.$$

Example 7. (CMC/2008) Given $\sec x + \tan x = \dfrac{22}{7}$, $\csc x + \cot x = \dfrac{m}{n}$, where $(m, n) = 1$. Find $m + n$.

Solution $\dfrac{1 + \sin x}{\cos x} = \dfrac{22}{7} \Rightarrow \dfrac{1 + \sin x + \cos x}{1 + \sin x - \cos x} = \dfrac{29}{15}$.

$$\because \quad \frac{1 + \sin x + \cos x}{1 + \sin x - \cos x} = \frac{2\cos^2 \frac{x}{2} + 2 \sin \frac{x}{2} \cos \frac{x}{2}}{2 \sin^2 \frac{x}{2} + 2 \sin \frac{x}{2} \cos \frac{x}{2}} = \cot \frac{x}{2} \quad \text{and}$$

$$\csc x + \cot x = \frac{1 + \cos x}{\sin x} = \frac{2 \cos^2 \frac{x}{2}}{2 \sin \frac{x}{2} \cos \frac{x}{2}} = \cot \frac{x}{2},$$

therefore $\dfrac{m}{n} = \dfrac{29}{15}$ and $m + n = 44$.

The identities in the following example are those involving three interior angles of a triangle, the results can be readily applied.

Example 8. Given that A, B, C are three inner angles of the $\triangle ABC$. Prove the following identities

$$(i) \qquad \sin A + \sin B + \sin C = 4\cos\frac{A}{2}\cos\frac{B}{2}\cos\frac{C}{2};$$

$$(ii) \qquad \cos A + \cos B + \cos C = 1 + 4\sin\frac{A}{2}\sin\frac{B}{2}\sin\frac{C}{2};$$

$$(iii) \qquad \tan A + \tan B + \tan C = \tan A \tan B \tan C;$$

$$(iv) \qquad \tan\frac{A}{2}\tan\frac{B}{2} + \tan\frac{B}{2}\tan\frac{C}{2} + \tan\frac{C}{2}\tan\frac{A}{2} = 1;$$

$$(v) \qquad \cot\frac{A}{2} + \cot\frac{B}{2} + \cot\frac{C}{2} = \cot\frac{A}{2}\cot\frac{B}{2}\cot\frac{C}{2};$$

$$(vi) \qquad \cot A \cot B + \cot B \cot C + \cot C \cot A = 1.$$

Solution

(i) $\sin A + \sin B + \sin C = 2\sin\dfrac{A+B}{2}\cos\dfrac{A-B}{2} + \sin(A+B)$

$$= 2\sin\frac{A+B}{2}\left[\cos\frac{A-B}{2} + \cos\frac{A+B}{2}\right]$$

$$= 4\sin\frac{A+B}{2}\cos\frac{A}{2}\cos\frac{B}{2} = 4\cos\frac{C}{2}\cos\frac{A}{2}\cos\frac{B}{2}.$$

(ii) $\cos A + \cos B + \cos C = 2\cos\dfrac{A+B}{2}\cos\dfrac{A-B}{2} - \cos(A+B)$

$$= 2\cos\frac{A+B}{2}\cos\frac{A-B}{2} - (2\cos^2\frac{A+B}{2} - 1)$$

$$= 1 + 2\cos\frac{A+B}{2}\left[\cos\frac{A-B}{2} - \cos\frac{A+B}{2}\right]$$

$$= 1 + 4\sin\frac{C}{2}\sin\frac{B}{2}\sin\frac{A}{2}.$$

(iii) $\tan A + \tan B + \tan C = \tan(A+B)(1 - \tan A \tan B) + \tan C$

$$= -\tan C(1 - \tan A \tan B) + \tan C = \tan A \tan B \tan C.$$

(iv) $\tan \dfrac{A}{2} \tan \dfrac{B}{2} + \tan \dfrac{B}{2} \tan \dfrac{C}{2} + \tan \dfrac{C}{2} \tan \dfrac{A}{2}$

$= \tan \dfrac{A}{2} \left(\tan \dfrac{B}{2} + \tan \dfrac{C}{2} \right) + \tan \dfrac{B}{2} \tan \dfrac{C}{2}$

$= \tan \dfrac{A}{2} \tan \left(\dfrac{B}{2} + \dfrac{C}{2} \right) \left(1 - \tan \dfrac{B}{2} \tan \dfrac{C}{2} \right) + \tan \dfrac{B}{2} \tan \dfrac{C}{2}$

$= \tan \dfrac{A}{2} \cot \dfrac{A}{2} \left(1 - \tan \dfrac{B}{2} \tan \dfrac{C}{2} \right) + \tan \dfrac{B}{2} \tan \dfrac{C}{2} = 1.$

(v) When the both sides of (iv) are divided by $\tan \dfrac{A}{2} \tan \dfrac{B}{2} \tan \dfrac{C}{2}$, the equality (v) is obtained at once.

(vi) When the both sides of (iii) are divided by $\tan A \tan B \tan C$, then (vi) is obtained at once.

Testing Questions   (A)

1. (SSSMO/2007) Evaluate $256 \sin 10° \sin 30° \sin 50° \sin 70°$.

2. (CMC/2008) Let $a_1, a_2, \ldots, a_n$ be the sequence of all irreducible proper fractions with the denominator 24, arranged in ascending order. Find the value of
$$\sum_{i=1}^{n} \cos(a_i \pi).$$

3. Prove that
$$\cos(\alpha + \beta) + \cos(\alpha + 2\beta) + \cdots + \cos(\alpha + n\beta) = \frac{\sin \frac{n\beta}{2} \cos(\alpha + \frac{n+1}{2}\beta)}{\sin \frac{1}{2}\beta}.$$

4. (SSSMO/2009) Find the value of
$$(\cot 25° - 1)(\cot 24° - 1)(\cot 23° - 1) \cdots (\cot 20° - 1).$$

5. (CROATIA/2004) Prove that $\tan^n 15° + \cot^n 15°$ must be an even positive integer for any positive integer n

6. Prove that for any positive integer n,
$$\tan \alpha \tan 2\alpha + \tan 2\alpha \tan 3\alpha + \cdots + \tan(n-1)\alpha \tan n\alpha = \frac{\tan n\alpha}{\tan \alpha} - n$$
where $\tan \alpha \neq 0$ and $\tan k\alpha \neq \pm\infty$ for $k = 1, 2, \ldots, n$.

7.   (CMC/2009) Given $0 < \alpha < \pi, \pi < \beta < 2\pi$. If the equality

$$\cos(x + \alpha) + \sin(x + \beta) + \sqrt{2}\cos x = 0$$

holds for any $x \in \mathbb{R}$, find the values of α and β.

8.   (SSSMO/2004) Find the value of $\sin^2 1° + \sin^2 2° + \sin^2 3° + \cdots + \sin^2 360°$.

9.   (AIME/2000) Find the smallest positive integer n such that

$$\frac{1}{\sin 45° \sin 46°} + \frac{1}{\sin 47° \sin 48°} + \cdots + \frac{1}{\sin 133° \sin 134°} = \frac{1}{\sin n°}.$$

10. (CMC/2009) Given $\sin \alpha + \sin \beta = \dfrac{\sqrt{6}}{3}, \cos \alpha + \cos \beta = \dfrac{\sqrt{3}}{3}$, find the value of $\cos^2 \dfrac{\alpha - \beta}{2}$.

Testing Questions   (B)

1.   (CMC/2009) Evaluate $\cos \dfrac{\pi}{15} - \cos \dfrac{2\pi}{15} - \cos \dfrac{4\pi}{15} + \cos \dfrac{7\pi}{15}$.

2.   (USAMO/TST/2005) Evaluate $\cos 36° - \cos 72°$.

3.   (SSSMO/2009) If $\dfrac{\cos 100°}{1 - 4\sin 25° \cos 25° \cos 50°} = \tan x°$, find x.

4.   (USAMO/TST/2005) Prove that

$$\frac{1}{\sin 1° \sin 2°} + \frac{1}{\sin 2° \sin 3°} + \cdots + \frac{1}{\sin 89° \sin 90°} = \frac{\cos 1°}{\sin^2 1°}.$$

5.   Prove that (i) $\tan \dfrac{\pi}{5} \tan \dfrac{2\pi}{5} = \sqrt{5}$;   (ii) $\tan^2 \dfrac{\pi}{5} + \tan^2 \dfrac{2\pi}{5} = 10$.

Lecture 8

Extreme Values of Functions and Mean Inequality

The following methods to find maximum or minimum values of a function are discussed in this lecture:

(I)   The extreme values of a function can be determined based on the *analysis of the tendency of its change*, since if an extreme value of $f(x)$ is taken at some point x_0, then the tendency of change of $f(x)$ must be changed at x_0. For this it is important to consider the graph of the function, like in the case that $f(x)$ contains absolute value signs.

(II) For quadratic function $y = ax^2 + bx + c$, the *completing the squares* method is powerful for getting its extreme values. However, when dealing with the quadratic function of multi-variables or conditional extreme value problems, completing the squares alone is not sufficient, it should be complemented with some other techniques, like canceling variables or substitution of variables, etc.

(III) To determining the range of one variable x_1 in a quadratic function of two variables x_1, x_2, it is often useful to consider x_1 as a constant at the moment and investigate the resulting quadratic equation of another variable x_2, then the non-negativity of its discriminant will yield an inequality of the variable x_1, from which the range of x_1 is obtained.

(IV) Instead of using the discriminant in the method (III), we can get an inequality in x_1 by using some inequalities on the other variables also. (cf. Example 5).

(V)   The *Mean inequality* (cf. Appendix B) can be used to enlarge or compress the value of a function considered, so that a constant upper bound or lower bound of the function is obtained, then the remaining work is to show the constants are reachable by the function.

(VI) By introducing *trigonometric transformations*, many functions can be converted to simple functions of six basic trigonometric functions, so that the techniques for dealing with trigonometric functions can be used for getting their extreme values.

Examples

Example 1. $f(x) = |x| + 2|x - 1| + |x - 2| + |x - 4| + |x - 6| + 2|x - 10|$, where $x \in \mathbb{R}$. Find the minimum value of f.

Solution   For any real numbers a and b with $a < b$, the function $g(x) = |x - a| + |x - b|, -\infty < x < +\infty$, can be written in the form

$$g(x) = \begin{cases} a + b - 2x & \text{if } x \leq a, \\ b - a & \text{if } a \leq x \leq b, \\ 2x - (a + b) & \text{if } b \leq x. \end{cases}$$

So $g(x)$ is decreasing on $(-\infty, a]$, constant $b - a$ on $[a, b]$, and increasing on $[b, +\infty)$.

Thus, g takes its minimum value $b - a$ when $a \leq x \leq b$.

Since $|x| + |x - 10|, |x - 1| + |x - 10|, |x - 1| + |x - 6|$, and $|x - 2| + |x - 4|$ take their minimum values on $[0, 10], [1, 10], [1, 6]$ and $[2, 4]$ respectively, therefore f takes its minimum value when $x \in [2, 4]$, and by letting $x = 2$, $f(2) = 2 + 2 + 0 + 2 + 4 + 16 = 26$ is the minimum value of f.

Example 2. (CHINA/2005) Given $|y| \leq 1$ and $2x + y = 1$, find the minimum value of $2x^2 + 16x + 3y^2$.

Solution $|y| \leq 1, 2x + y = 1 \Leftrightarrow 2x = 1 - y, -1 \leq y \leq 1 \Rightarrow 0 \leq x \leq 1$. Therefore

$$\begin{aligned} 2x^2 + 16x + 3y^2 &= 2x^2 + 16x + 3(1 - 2x)^2 = 14x^2 + 4x + 3 \\ &= 14\left(x + \frac{1}{7}\right)^2 + \frac{19}{7}. \end{aligned}$$

Thus, the minimum value of the given expression is taken when $x = 0$, so it is 3.

Example 3. (CMC/2009) Let x, y be real numbers satisfying $2x + y \geq 1$. Find the minimum value of the function of two variables $u = x^2 + 4x + y^2 - 2y$.

Solution   If complete squares for x and y separately, then the condition $2x + y \geq 1$ cannot be satisfied. Now let $z = 2x + y$, then $z \geq 1$, $y = z - 2x$ and

$$\begin{aligned} u &= x^2 + 4x + (z - 2x)^2 - 2(z - 2x) = 5x^2 - 4zx + 8x + z^2 - 2z \\ &= 5\left(x^2 - \frac{4}{5}(z - 2)x + \frac{4}{25}(z - 2)^2\right) + \left(z^2 - 2z - \frac{4}{5}(z - 2)^2\right) \\ &= 5\left(x - \frac{2}{5}(z - 2)\right)^2 + \frac{z^2 + 6z - 16}{5} \geq \frac{1 + 6 - 16}{5} = -\frac{9}{5}. \end{aligned}$$

The equality holds when $z = 1, x = \frac{2}{5}(z - 2) = -\frac{2}{5}$. Thus, $u_{\min} = -\frac{9}{5}$.

Example 4. Find the maximum value and minimum value of $y = \dfrac{3x^2 - 2x + 2}{x^2 + 2x + 2}$.

Solution Consider the value of y is taken in its range, so y is considered as a constant. By moving the denominator to the left hand side, then an equation in x is obtained

$$yx^2 + 2yx + 2y = 3x^2 + 2x + 2 \Leftrightarrow (y-3)x^2 + 2(y-1)x + (2y-2) = 0.$$

The equation must have real roots, so its discriminant is non-negative, i.e., if $y \neq 3$, then

$$(y-1)^2 - (y-3)(2y-2) \geq 0 \Rightarrow y^2 - 6y + 5 \leq 0$$
$$\Rightarrow (y-1)(y-5) \leq 0 \Rightarrow 1 \leq y \leq 5.$$

When $y = 3$, then $x = -1$. Thus, $y_{min} = 1$ and $y_{max} = 5$.

Example 5. (USAMO/1978) Given that the real numbers a, b, c, d, e satisfy $a + b + c + d + e = 8$ and $a^2 + b^2 + c^2 + d^2 + e^2 = 16$. Find the maximum value of e.

Solution First of all we prove the following inequality: For any real a, b, c, d

$$4(a^2 + b^2 + c^2 + d^2) \geq (a + b + c + d)^2,$$

and the equality holds if and only if $a = b = c = d$. In fact,

$$4(a^2 + b^2 + c^2 + d^2) - (a + b + c + d)^2$$
$$= 3(a^2 + b^2 + c^2 + d^2) - 2(ab + ac + ad + bc + bd + cd)$$
$$= (a^2 - 2ab + b^2) + (a^2 - 2ac + c^2) + (a^2 - 2ad + d^2) + (b^2 - 2bc + c^2)$$
$$+ (b^2 - 2bd + d^2) + (c^2 - 2cd + d^2)$$
$$= (a-b)^2 + (a-c)^2 + (a-d)^2 + (b-c)^2 + (b-d)^2 + (c-d)^2 \geq 0.$$

Thus, $a + b + c + d = (8 - e)$ and $a^2 + b^2 + c^2 + d^2 = 16 - e^2$ gives the inequality in e

$$(8 - e)^2 \leq 4(16 - e^2),$$
$$5e^2 - 16e \leq 0,$$
$$5e\left(e - \frac{16}{5}\right) \leq 0,$$
$$\therefore 0 \leq e \leq \frac{16}{5}.$$

It's easy to see that $e = \dfrac{16}{5}$ when $a = b = c = d = \dfrac{6}{5}$. Thus $e_{max} = \dfrac{16}{5}$.

Example 6. (CROATIA/2004) Given $a > 0$. Find the minimum value of the function $f(x) = x^5 + \dfrac{a}{x}$ ($x > 0$).

Solution   By AM-GM inequality,

$$f(x) = x^5 + \frac{a}{5x} + \frac{a}{5x} + \frac{a}{5x} + \frac{a}{5x} + \frac{a}{5x} \geq 6 \sqrt[6]{x^5 \cdot \left(\frac{a}{5x}\right)^5} = 6 \sqrt[6]{\left(\frac{a}{5}\right)^5},$$

the equality holds if and only if $x^5 = \dfrac{a}{5x}$ i.e. $x = \sqrt[6]{\dfrac{a}{5}}$. Thus the minimum value of the function is

$$f\left(\sqrt[6]{\frac{a}{5}}\right) = 6 \sqrt[6]{\left(\frac{a}{5}\right)^5}.$$

Example 7. (SMO/2003) Find the maximum value of

$$\frac{xyz}{(1 + 5x)(4x + 3y)(5y + 6z)(z + 18)}$$

as x, y and z range over the set of all positive real numbers. Justify your answer.

Solution   Let $I = \dfrac{xyz}{(1 + 5x)(4x + 3y)(5y + 6z)(z + 18)}$, then

$$I = \frac{xyz}{(1 + 5x)(4x + 3y)(5y + 6z)(z + 18)} = \frac{1}{20(1 + \alpha)(1 + \beta)(1 + \gamma)(1 + \delta)},$$

where

$$\alpha = 5x, \quad \beta = \frac{3y}{4x}, \quad \gamma = \frac{6z}{5y}, \quad \delta = \frac{18}{z},$$

hence $\alpha, \beta, \gamma, \delta$ are all positive with $\alpha \cdot \beta \cdot \gamma \cdot \delta = 81 = 3^4$. Then the mean inequality gives

$$\begin{aligned}
&(1 + \alpha)(1 + \beta)(1 + \gamma)(1 + \delta) \\
&= 1 + (\alpha + \beta + \gamma + \delta) + (\alpha\beta + \alpha\gamma + \alpha\delta + \beta\gamma + \beta\delta + \gamma\delta) \\
&\quad + (\alpha\beta\gamma + \alpha\beta\delta + \alpha\gamma\delta + \beta\gamma\delta) + \alpha\beta\gamma\delta \\
&\geq 1 + 4(\alpha \cdot \beta \cdot \gamma \cdot \delta)^{\frac{1}{4}} + 6(\alpha \cdot \beta \cdot \gamma \cdot \delta)^{\frac{2}{6}} + 4(\alpha \cdot \beta \cdot \gamma \cdot \delta)^{\frac{3}{4}} + \alpha \cdot \beta \cdot \gamma \cdot \delta \\
&= 256.
\end{aligned}$$

Hence $I \leq \dfrac{1}{5120}$. Since $I = \dfrac{1}{5120}$ when $\alpha = \beta = \gamma = \delta = 3$, i.e., when $x = \dfrac{3}{5}, y = \dfrac{12}{5}, z = 6$, it proves that $I_{\max} = \dfrac{1}{5120}$.

Example 8. (CMC/2009) Given that the lengths of three sides of $\triangle ABC$ are $3, 4$ and 5. P is a point variable in the interior of $\triangle ABC$ (not on its boundary). Find the maximum value of product of distances from P to the three sides AB, BC, CA.

Solution   Let $a = BC = 3, b = CA = 4, c = AB = 5$, then $\triangle ABC$ is a right triangle with $\angle C = 90°$. Use h_a, h_b, h_c to denote the distances from P to BC, CA, AB respectively, then

$$ah_a + bh_b + ch_c = 2[ABC] = 12,$$

therefore, by the mean inequality,

$$
\begin{aligned}
h_a h_b h_c &= \frac{(ah_a)(bh_b)(ch_c)}{abc} \leq \frac{1}{abc}\left(\frac{ah_a + bh_b + ch_c}{3}\right)^3 \\
&= \frac{4^3}{60} = \frac{16}{15}.
\end{aligned}
$$

The equality holds when $ah_a = bh_b = ch_c = 4$, i.e., $[PAB] = [PBC] = [PCA]$, which means that P is the center of gravity of $\triangle ABC$.

Thus, the maximum value of $h_a h_b h_c$ is $\dfrac{16}{15}$.

Example 9. (CMC/2008) Find the maximum value of $z = \dfrac{x + y - 3}{x - y + 1}$, where x, y satisfy $(x - 3)^2 + 4(y - 1)^2 = 4$.

Solution   Let $x = 3 + 2\cos\theta$, $y = 1 + \sin\theta$, then

$$z = \frac{x + y - 3}{x - y + 1} = \frac{2\cos\theta + \sin\theta + 1}{2\cos\theta - \sin\theta + 3}.$$

When z has taken a value in its range, move the denominator to the left hand side, then

$$(2z - 2)\cos\theta - (z + 1)\sin\theta + 3z - 1 = 0.$$

Using $\cos\theta = \dfrac{1 - \tan^2\frac{\theta}{2}}{1 + \tan^2\frac{\theta}{2}}$, $\sin\theta = \dfrac{2\tan\frac{\theta}{2}}{1 + \tan^2\frac{\theta}{2}}$, then

$$(z + 1)\tan^2\frac{\theta}{2} - 2(z + 1)\tan\frac{\theta}{2} + 5z - 3 = 0.$$

When $z \neq -1$, then the discriminant of the quadratic equation in $\tan\frac{\theta}{2}$ is nonnegative, so

$$\frac{1}{4}\Delta = (z + 1)^2 - (z + 1)(5z - 3) \geq 0 \Rightarrow z^2 - 1 \leq 0 \Rightarrow -1 < z \leq 1.$$

When $z = -1$ then $x + y - 3 = -(x - y + 1)$, so $x = 1$ or $\cos\theta = -1$, $y = 1$. When $z = 1$, then $\tan\frac{\theta}{2} = 1$, so $x = 3, y = 2$. Thus,

$$z_{\max} = 1, \qquad z_{\min} = -1.$$

Example 10. (CROATIA/2005) If k, m, n are positive integers with $\dfrac{1}{k} + \dfrac{1}{m} + \dfrac{1}{n} < 1$, find the maximum possible value of $\dfrac{1}{k} + \dfrac{1}{m} + \dfrac{1}{n}$.

Solution   Let $M(k, m, n) = \dfrac{1}{k} + \dfrac{1}{m} + \dfrac{1}{n}$. WLOG, we may assume $k \leq m \leq n$. Below we discuss three cases: $k = 2, k = 3, k \geq 4$.

(i)   When $k = 2$, since $\dfrac{1}{2} + \dfrac{1}{m} + \dfrac{1}{n} < 1$, then $m > 2$.

   (1)   If $m = 3$, then $\dfrac{1}{n} < \dfrac{1}{6}$ i.e. $n > 6$. From $n = 7$, we obtain
$$\max\{M\} = \frac{1}{2} + \frac{1}{3} + \frac{1}{7} = \frac{41}{42}.$$

   (2)   If $m = 4$, then $\dfrac{1}{n} < \dfrac{1}{4}$ i.e. $n > 4$. From $n = 5$, we obtain
$$\max\{M\} = \frac{1}{2} + \frac{1}{4} + \frac{1}{5} = \frac{19}{20}.$$

   (3)   If $m > 4$, from $\dfrac{1}{2} + \dfrac{1}{m} + \dfrac{1}{n} < \dfrac{1}{2} + \dfrac{1}{4} + \dfrac{1}{4} = 1$ and $5 \leq m \leq n$, we
obtain
$$\max\{M\} = \frac{1}{2} + \frac{1}{5} + \frac{1}{5} = \frac{9}{10}.$$

(ii)   When $k = 3$, we consider the cases $m = 3$ and $m > 3$.

   (1)   If $m = 3$, then $\dfrac{1}{n} < \dfrac{1}{3}$ i.e. $n > 3$. From $n = 4$, we obtain $\max\{M\} = \dfrac{11}{12}$.

   (2)   If $m > 3$, from $m = n = 4$ we have $\max\{M\} = \dfrac{5}{6}$.

(iii)   When $k \geq 4$, from $\dfrac{1}{k} + \dfrac{1}{m} + \dfrac{1}{n} \leq \dfrac{1}{4} + \dfrac{1}{4} + \dfrac{1}{4} < 1$ we obtain $\max\{M\} = \dfrac{3}{4}$.
Thus, $\max\{M\} = \dfrac{41}{42}$.

Testing Questions   (A)

1. If the minimum value of $f(x) = |x + 1| + |ax + 1|, x \in \mathbb{R}$ is $\dfrac{3}{2}$, find the possible values of the real parameter a.

2. Find the minimum value of the function
    $$y = (x^2 + 4x + 5)(x^2 + 4x + 1) + 3x^2 + 12x + 5, \quad x \in \mathbb{R}.$$

3. (CMC/2009) For natural number n and function $f(x) = \dfrac{x^2 + n}{x^2 + x + 1}$, let its maximum value be a_n and it minimum value be b_n, find $a_n - b_n$.

4. (CMC/2009) Given that $A(0, a)$ $(a > 1)$ is on the y-axis, $M(x, y)$ is a moving point which moves on the curve $y = \left|\dfrac{1}{2}x^2 - 1\right|$. Find the minimum value of the distance $|AM|$, in terms of a.

5. (CMC/2009) a, b are positive constants. Find the minimum value of $f(x) = \dfrac{a^2}{x} + \dfrac{b^2}{1 - x}, 0 < x < 1$.

6. (CMC/2008) Find the minimum value of the function $f(x) = \dfrac{5 - 4x + x^2}{2 - x}$, where $-\infty < x < 2$.

7. (HUNGARY/2003) Given that the non-negative numbers x, y, z satisfy $x^2 + y^2 + z^2 + x + 2y + 3z = \dfrac{13}{4}$.
    (i)   Find the maximum value of $x + y + z$;
    (ii) Prove that $x + y + z \geq \dfrac{\sqrt{22} - 3}{2}$.

8. (CMC/2009) In a tetrahedron $PABC$, $\angle APB = \angle BPC = \angle CPA = 90°$, and the sum of all edges is S, find the maximum volume of such tetrahedra.

9. (JAPAN/2005) Given that a, b are real numbers such that $a + b = 17$, find the minimum value of $2^a + 4^b$.

10. (IMO/1976) In a convex quadrilateral of area 64 cm², the sum of the lengths of a diagonal and a pair of opposite sides is $16\sqrt{2}$ cm. Suppose the length of the other diagonal is x cm. Find the value of x.

Testing Questions   (B)

1. (CMC/2009) If $f(x) = x^2 - 2x - |x - 1 - a| - |x - 2| + 4$, $x \in \mathbb{R}$ is always non-negative, find the minimum value of the real parameter a.

2. (CMC/2009) Given the function $f(x) = 3ax^2 - 2(a + b)x + b$, where $a > 0, b \in \mathbb{R}$. Prove that

$$|f(x)| \leq \max\{f(0), f(1)\}, \quad \text{if } 0 \leq x \leq 1.$$

3. (CMC/2009) $x_1, x_2, \ldots, x_{2010} > 0$ and $\displaystyle\sum_{i=1}^{2010} x_i^{2009} = 1$. Find

$$\min \left\{ \sum_{i=1}^{2010} \frac{x_i^{2008}}{1 - x_i^{2009}} \right\},$$

and prove your result.

4. (USAMO/2002) Find the maximum value of

$$S = (1 - x_1)(1 - y_1) + (1 - x_2)(1 - y_2)$$

if $x_1^2 + x_2^2 = y_1^2 + y_2^2 = c^2$.

5. (IRE/2003) Given $a, b > 0$. Find the maximum positive integer c such that for any positive real number x,

$$c \leq \max \left\{ ax + \frac{1}{ax}, bx + \frac{1}{bx} \right\}.$$

Lecture 9

Extreme Value Problems in Trigonometry

Basic Methods for Solving Trigonometric Inequalities and Trigonometric Extreme Value Problems

1. Make use of the boundedness and monotonic intervals of the six basic trigonometric functions.
2. Make use of trigonometric identities to simplify or convert the given inequality or trigonometric function.
3. For finding the extreme values of a given trigonometric function, the principle of extremum property is often useful.
4. Apply some basic inequalities involving a triangle, like

$$|a \cos x + b \sin x| \le \sqrt{a^2 + b^2}.$$

(Refer to Appendix C for more).
5. Sometimes, by using substitutions of variables or expressions, a trigonometric inequality or a trigonometric extreme value problem can be converted to an algebraic inequality or an algebraic extreme value problem.

Examples

Example 1. (RUSMO/2004) Let a, b, c be positive numbers, satisfying $a + b + c = \dfrac{\pi}{2}$, prove that

$$\cos a + \cos b + \cos c > \sin a + \sin b + \sin c.$$

Solution $a + b < \dfrac{\pi}{2} \Rightarrow a < \dfrac{\pi}{2} - b \Rightarrow \cos a > \cos\left(\dfrac{\pi}{2} - b\right) = \sin b$ since $\cos\theta$ is decreasing on the first quadrant.

Similarly, $\cos b > \sin c$ and $\cos c > \sin a$. Adding them up gives the conclusion at once.

61

Example 2. (CMC/2008) Given the function $f(x) = \dfrac{\sin(x + 45°)}{\sin(x + 60°)}, x \in [0°, 90°]$, find product of the maximum value and minimum value of $f(x)$.

Solutions By the substitution $z = x + 45°, x \in [0, 90°]$,

$$f(x) = f(z - 45°) = g(z) = \frac{\sin z}{\sin(z + 15°)}, \quad z \in [45°, 135°].$$

For any $45° \leq z \leq 135°$ and $0° < \Delta z \leq 1°$,

$$
\begin{aligned}
g(z + \Delta z) - g(z) &= \frac{\sin(z + \Delta z)}{\sin(z + \Delta z + 15°)} - \frac{\sin z}{\sin(z + 15°)} \\
&= \frac{\sin(z + 15°)\sin(z + \Delta z) - \sin z \sin(z + \Delta z + 15°)}{\sin(z + \Delta z + 15°)\sin(z + 15°)}.
\end{aligned}
$$

The denominator is always positive. For the numerator,

$$
\begin{aligned}
&\sin(z + 15°)\sin(z + \Delta z) - \sin z \sin(z + \Delta z + 15°) \\
&= \sin(z + 15°)[\sin z \cos \Delta z + \cos z \sin \Delta z] \\
&\quad - \sin z[\sin(z + 15°)\cos \Delta z + \cos(z + 15°)\sin \Delta z] \\
&= \sin \Delta z[\sin(z + 15°)\cos z - \sin z \cos(z + 15°)] \\
&= \sin \Delta z \sin 1° > 0,
\end{aligned}
$$

therefore g is increasing strictly on its domain, and so is f. Thus,

$$\max\{f(x)\} = \frac{\sin 135°}{\sin 150°} = \sqrt{2}, \quad \min\{f(x)\} = \frac{\sin 45°}{\sin 60°} = \frac{\sqrt{2}}{\sqrt{3}},$$

and their product is $\dfrac{2\sqrt{3}}{3}$.

Example 3. (CMC/2008) Find the maximum value of the function

$$y = \cos^3 x + \sin^2 x - \cos x, \qquad x \in \mathbb{R}.$$

Solution By using the trigonometric identities and then using the AM-GM inequality,

$$
\begin{aligned}
y &= \sin^2 x - \cos x(1 - \cos^2 x) = \sin^2 x(1 - \cos x) \\
&= 2 \sin^2 x \sin^2 \frac{x}{2} = 8 \sin^4 \frac{x}{2} \cos^2 \frac{x}{2} \\
&= 4 \sin^2 \frac{x}{2} \cdot \sin^2 \frac{x}{2} \cdot 2 \cos^2 \frac{x}{2} \\
&\leq 4\left(\frac{\sin^2 \frac{x}{2} + \sin^2 \frac{x}{2} + 2\cos^2 \frac{x}{2}}{3}\right)^3 = \frac{32}{27}.
\end{aligned}
$$

The equality holds if and only if $\sin^2 \frac{x}{2} = 2\cos^2 \frac{x}{2}$, namely $\tan \frac{x}{2} = \pm\sqrt{2}$. Thus,

$$y_{max} = \frac{32}{27}.$$

Example 4. (CMC/2008) Find the maximum value of the function

$$y = \left[\sin\left(\frac{\pi}{4} + x\right) - \sin\left(\frac{\pi}{4} - x\right)\right]\sin\left(\frac{\pi}{3} + x\right),$$

and find the set of corresponding values of x where f takes its maximum value.

Solution Simplify the function and then use the R-formula,

$$
\begin{aligned}
y &= \left[\sin\left(\frac{\pi}{4} + x\right) - \sin\left(\frac{\pi}{4} - x\right)\right]\sin\left(\frac{\pi}{3} + x\right) \\
&= \left[\frac{\sqrt{2}}{2}(\sin x + \cos x) - \frac{\sqrt{2}}{2}(\cos x - \sin x)\right]\left(\frac{\sqrt{3}}{2}\cos x + \frac{1}{2}\sin x\right) \\
&= \sqrt{2}\sin x \left(\frac{\sqrt{3}}{2}\cos x + \frac{1}{2}\sin x\right) = \frac{\sqrt{6}}{2}\sin x \cos x + \frac{\sqrt{2}}{2}\sin^2 x \\
&= \frac{\sqrt{6}}{4}\sin 2x + \frac{\sqrt{2}}{4}(1 - \cos 2x) = \frac{\sqrt{2}}{2}\left(\frac{\sqrt{3}}{2}\sin 2x - \frac{1}{2}\cos 2x\right) + \frac{\sqrt{2}}{4} \\
&= \frac{\sqrt{2}}{2}\sin\left(2x - \frac{\pi}{6}\right) + \frac{\sqrt{2}}{4},
\end{aligned}
$$

therefore $y_{max} = \dfrac{\sqrt{2}}{2} + \dfrac{\sqrt{2}}{4} = \dfrac{3\sqrt{2}}{4}$, and the corresponding set of x is

$$\left\{x : x = k\pi + \frac{\pi}{3},\ k \in \mathbb{Z}\right\}.$$

Example 5. (CMC/2010) Given that the three sides of $\triangle ABC$ have distinct lengths, the angle bisectors of $\angle A$, $\angle B$, $\angle C$ intersects the perpendicular bisectors of BC, CA, AB at D, E, F respectively. Prove that the area of $\triangle ABC$ is less than that of $\triangle DEF$.

Solution From the assumptions it is easy to show that A, B, C, D, E, F are concyclic. We may assume that the circumradius of $\triangle ABC$ is 1.
Since

$$[ABC] = \frac{abc}{4R} = 2\sin A \sin B \sin C = \frac{1}{2}(\sin 2A + \sin 2B + \sin 2C)$$

and

$$[DEF] = \frac{1}{2}(\sin(A+B) + \sin(B+C) + \sin(C+A)) = \frac{1}{2}(\sin A + \sin B + \sin C),$$

from
$$\sin 2A + \sin 2B + \sin 2C$$
$$= \frac{1}{2}(\sin 2A + \sin 2B) + \frac{1}{2}(\sin 2B + \sin 2C) + \frac{1}{2}(\sin 2C + \sin 2A)$$
$$= \sin(A + B)\cos(A - B) + \sin(B + C)\cos(B - C) + \sin(C + A)\cos(C - A)$$
$$< \sin(A + B) + \sin(B + C) + \sin(C + A) = \sin A + \sin B + \sin C,$$

it follows that $[ABC] < [DEF]$, as desired.

Example 6. (INDIA/2007) The three sides of $\triangle ABC$ have lengths a, b, c, and the corresponding angle bisectors have lengths w_a, w_b, w_c respectively. If the circumradius of the $\triangle ABC$ is R, prove that

$$\frac{a^2 + b^2}{w_c} + \frac{b^2 + c^2}{w_a} + \frac{c^2 + a^2}{w_b} > 4R.$$

Solution   By Stewart's theorem,

$$w_a = \sqrt{\frac{bc}{(b + c)^2}[(b + c)^2 - a^2]} = \frac{bc}{b + c}\sqrt{2(1 + \cos A)} = \frac{2bc}{b + c}\cos\frac{A}{2}$$

and similarly,

$$w_b = \frac{2ca}{c + a}\cos\frac{B}{2}, \qquad w_c = \frac{2ab}{a + b}\cos\frac{C}{2}.$$

Therefore

$$\frac{a^2 + b^2}{w_c} + \frac{b^2 + c^2}{w_a} + \frac{c^2 + a^2}{w_b} > 4R$$

$$\Leftrightarrow \frac{(b^2 + c^2)(b + c)}{4Rbc\cos\frac{A}{2}} + \frac{(c^2 + a^2)(c + a)}{4Rca\cos\frac{B}{2}} + \frac{(a^2 + b^2)(a + b)}{4Rab\cos\frac{C}{2}} > 2$$

$$\Leftrightarrow \frac{(b^2 + c^2)(b + c)\sin\frac{A}{2}}{2abc} + \frac{(c^2 + a^2)(c + a)\sin\frac{B}{2}}{2abc} + \frac{(a^2 + b^2)(a + b)\sin\frac{C}{2}}{2abc}$$
$$> 1.$$

Since $b^2 + c^2 \geq 2bc$ and $b + c > a$ and the similar inequalities, it suffices to show that
$$\sin\frac{A}{2} + \sin\frac{B}{2} + \sin\frac{C}{2} > 1.$$

Since $\dfrac{\pi - A}{2} + \dfrac{\pi - B}{2} + \dfrac{\pi - C}{2} = \pi$, therefore

$$\cos\frac{\pi - A}{2} + \cos\frac{\pi - B}{2} + \cos\frac{\pi - C}{2} = 1 + 4\sin\frac{A}{2}\sin\frac{B}{2}\sin\frac{C}{2} > 1,$$

thus,

$$\sin\frac{A}{2} + \sin\frac{B}{2} + \sin\frac{C}{2} = \cos\frac{\pi - A}{2} + \cos\frac{\pi - B}{2} + \cos\frac{\pi - C}{2} > 1.$$

Example 7. (CWMO/2008) Let A, B, C be three angles in $\left(0, \frac{\pi}{2}\right)$, satisfying

$$\sin 2A + \sin 2B + \sin 2C = 4\cos A \cos B \cos C,$$

find the maximum value of $\cos A \cos B \cos C$.

Solution Let $x = \cos^2 A, y = \cos^2 B, z = \cos^2 C$, then $x, y, z \in (0, 1)$, satisfying

$$\sqrt{x(1-x)} + \sqrt{y(1-y)} + \sqrt{z(1-z)} = 2\sqrt{xyz}.$$

By the AM-GM inequality,

$$\begin{aligned}
2\sqrt{xyz} &= \frac{1}{\sqrt{3}}[\sqrt{x(3-3x)} + \sqrt{y(3-3y)} + \sqrt{z(3-3z)}] \\
&\leq \frac{1}{\sqrt{3}}\left[\frac{x+(3-3x)}{2} + \frac{y+(3-3y)}{2} + \frac{z+(3-3z)}{2}\right] \\
&= \frac{3\sqrt{3}}{2} - \frac{1}{\sqrt{3}}(x+y+z) \leq \frac{3\sqrt{3}}{2} - \sqrt{3} \cdot \sqrt[3]{xyz}.
\end{aligned}$$

Let $\sqrt[6]{xyz} = p$, then

$$2p^3 \leq \frac{3\sqrt{3}}{2} - \sqrt{3}p^2 \Leftrightarrow 4p^3 + 2\sqrt{3}p^2 - 3\sqrt{3} \leq 0$$
$$\Leftrightarrow (2p - \sqrt{3})(2p^2 + 2\sqrt{3}p + 3) \leq 0 \Leftrightarrow 2p - \sqrt{3} \leq 0 \Leftrightarrow p \leq \frac{\sqrt{3}}{2}.$$

Thus $\cos A \cos B \cos C = \sqrt{xyz} = p^3 \leq \frac{3\sqrt{3}}{8}$, the equality holds when $x = y = z = \frac{3}{4}$, so

$$\max\{\cos A \cos B \cos C\} = \frac{3\sqrt{3}}{8}.$$

Example 8. (CMC/2009) In $\triangle ABC$,

$$(\sqrt{3}\sin B - \cos B)(\sqrt{3}\sin C - \cos C) = 4\cos B \cos C$$

and $AB + AC = 4$. Find the range of BC.

Solution   Dividing both sides of the given equality by $\cos B \cos C$ gives

$$(\sqrt{3}\tan B - 1)(\sqrt{3}\tan C - 1) = 4,$$
$$3\tan B \tan C - \sqrt{3}\tan B - \sqrt{3}\tan C + 1 = 4,$$
$$\sqrt{3}(\tan B + \tan C) = 3(\tan B \tan C - 1),$$
$$\tan(B + C) = -\sqrt{3}.$$

Since $0 < B + C < \pi$, so $B + C = \dfrac{2\pi}{3}, A = \dfrac{\pi}{3}$. $AB + BC = 4$ gives $AB = 4 - AC$. By the cosine rule,

$$\begin{aligned}
BC^2 &= AB^2 + AC^2 - 2AB \cdot AC \cos A \\
&= (4 - AC)^2 + AC^2 - (4 - AC) \cdot AC \\
&= 3AC^2 - 12AC + 16 = 3(AC - 2)^2 + 4 \geq 4.
\end{aligned}$$

Therefore $2 \leq BC < AB + BC = 4$, the range of BC is $[2, 4)$.

Example 9. (CMC/2009) Let $y = \sin x + \cos x + \tan x + \cot x + \sec x + \csc x$. Find the minimum value of $|y|$.

Solutions Let $t = \sin x + \cos x$, where $|t| \leq \sqrt{2}, |t| \neq 1$, then $\sin x \cos x = \dfrac{1}{2}(t^2 - 1)$ and

$$\begin{aligned}
y &= \sin x + \cos x + \frac{1}{\sin x \cos x} + \frac{\sin x + \cos x}{\sin x \cos x} \\
&= t + \frac{2}{t^2 - 1} + \frac{2t}{t^2 - 1} = t + \frac{2}{t - 1} \\
&= \left(t - 1 + \frac{2}{t - 1}\right) + 1.
\end{aligned}$$

The function $f(u) = u + \frac{1}{u}, u > 0$ is decreasing on $(0, 1]$. When $\sqrt{2} \geq t > 1$ and $u = \dfrac{t - 1}{\sqrt{2}}$, then $0 < u < \dfrac{\sqrt{2} - 1}{\sqrt{2}}$ and

$$y = \sqrt{2}f(u) + 1 \geq \left(\sqrt{2} - 1 + \frac{2}{\sqrt{2} - 1}\right) + 1 = 3\sqrt{2} + 2.$$

When $-\sqrt{2} \leq t < 1$, then $-y = 1 - t + \frac{2}{1 - t} - 1$ which takes its minimum value when $\dfrac{1 - t}{\sqrt{2}} = 1$ i.e. $t = 1 - \sqrt{2}$, and the minimum value is $2\sqrt{2} - 1$.

Thus, $|y|_{\min} = 2\sqrt{2} - 1$.

Testing Questions   (A)

1. (CMC/2008) Given $\alpha \in \mathbb{R}$ and

$$f(x) = \frac{\sin(x + \alpha) + \sin(x - \alpha) - 4\sin x}{\cos(x + \alpha) + \cos(x - \alpha) - 4\cos x}, \quad x \in \left(0, \frac{\pi}{4}\right),$$

    find the range of f.

2. (CMC/2008) Given that the minimum period of the function $f(x) = \cos^2 \theta x + \cos \theta x \sin \theta x$ is $\frac{\pi}{2}$, find the maximum value of $\theta f(x)$.

3. (CMC/2010) Given that the minimum value of the function $y = (a\cos^2 x - 3)\sin x$ is -3, find the range of the real number a.

4. (CMC/2010) Let a, b, c be the lengths of three sides of the right triangle ABC, where c be the length of the hypotenuse AB. Find the range of y given by

$$y = \frac{a^3 + b^3 + c^3}{c(a + b + c)^2}.$$

5. (THAILAND/2004) Let $n > 2, \alpha_1, \alpha_2, \ldots, \alpha_n \in \mathbb{R}$. Find the minimum value of

$$\sum_{1 \leq i < j \leq n} \cos^2(\alpha_i - \alpha_j).$$

6. (CHINA/2005) Prove that in an acute triangle ABC, if R and r are the circumradius and inradius of $\triangle ABC$, then

$$\frac{abc}{\sqrt{2(a^2 + b^2)(b^2 + c^2)(c^2 + a^2)}} \geq \frac{r}{2R}.$$

7. (CNMO/2007) Let $\alpha, \beta \in \left(0, \frac{\pi}{2}\right)$, find the maximum value of

$$A = \frac{\left(1 - \sqrt{\tan \frac{\alpha}{2} \tan \frac{\beta}{2}}\right)^2}{\cot \alpha + \cot \beta}.$$

8. (CMC/2008) Find the minimum value of $\dfrac{\sin^3 \alpha}{\cos \alpha} + \dfrac{\cos^3 \alpha}{\sin \alpha}, \alpha \in \left(0, \frac{\pi}{2}\right)$.

    (A) $\dfrac{27}{64}$;   (B) $\dfrac{3\sqrt{2}}{5}$;   (C) 1;   (D) $\dfrac{5\sqrt{3}}{6}$.

9.   (CROATIA/2007) In an acute triangle ABC, A_1, B_1, C_1 are the midpoints of BC, CA, AB respectively. O is the circumcenter and the circumradius is 1. Prove that
$$\frac{1}{OA_1} + \frac{1}{OB_1} + \frac{1}{OC_1} \geq 6.$$

10. (CMC/2008) Given that $\cos x + \cos y = 1$, Then the range of $\sin x - \sin y$ is

   (A) $[-1, 1]$;   (B) $[-2, 2]$;   (C) $[0, \sqrt{3}]$;   (D) $[-\sqrt{3}, \sqrt{3}]$.

Testing Questions   (B)

1.   (RUSMO/2004) For what positive integer n does the inequality
$$\sin n\alpha + \sin n\beta + \sin n\gamma < 0$$
hold always for any three interior angles α, β, γ of any acute triangle?

2.   (CROATIA/2004) Given the real numbers x, y, z. Prove the inequality
$$\sin^2 x \cos y + \sin^2 y \cdot \cos z + \sin^2 z \cdot \cos x < \frac{3}{2}.$$

3.   (CMC/2010) For any $x_i \geq 0, i = 1, 2, \ldots, n$, if let $x_{n+1} = x_1$, prove that
$$\sum_{k=1}^{n} \sqrt{\frac{1}{(x_k + 1)^2} + \frac{x_{k+1}^2}{(x_{k+1} + 1)^2}} \geq \frac{n}{\sqrt{2}}.$$

4.   (SMO/2008) Let $0 < a, b < \frac{\pi}{2}$. Show that
$$\frac{5}{\cos^2 a} + \frac{5}{\sin^2 a \sin^2 b \cos^2 b} \geq 27 \cos a + 36 \sin a.$$

5.   (ROMANIA/2006) Prove that for $a, b \in \left(0, \frac{\pi}{4}\right)$ and $n \in \mathbb{N}$, we have
$$\frac{\sin^n a + \sin^n b}{(\sin a + \sin b)^n} \geq \frac{\sin^n 2a + \sin^n 2b}{(\sin 2a + \sin 2b)^n}.$$

Lecture 10

Fundamental Properties of Circles

I.   A circle is symmetric with respect to its center, and also axisymmetric with respect to its each diameter. Therefore any equal angles at center are subtended by equal chord and equal arcs, and *vice versa*.

II.   A diameter bisects a chord and its subtended arc if and only if the diameter is perpendicular to the chord. By use of this property, it is easy to prove that two arcs between two parallel chords are equal.

III. An angle at circle is equal to half of the angle at center subtended by the same chord or the same arc. An angle at center can be measured by its subtended arc, in degrees. Therefore an angle at circle can be measured, in degrees, by half of its subtended arc.
      In particular, the angle at circle subtended by a diameter must be 90°.

IV.   For a circle of a radius R and a chord with a length $2l$, the *perpendicular distance of the center from the chord*, denoted by d, is given by

$$d = \sqrt{R^2 - l^2}.$$

Examples

Example 1. (SSSMO/2005) In the diagram, P, Q and R are three points on the circle whose centre is O. The lines PO and QR are produced to meet at S. Suppose that $RS = OP$, and $\angle PSQ = 12°$ and $\angle POQ = x°$. Find the value of x.

(A) 36   (B) 42   (C) 48
(D) 54   (E) 60

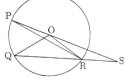

Solution   Connect PR, OR. Then $OR = RS$ implies that $\angle SOR = \angle PSR =$

69

12°, hence

$$\angle SPR = 6°.$$

Since $\dfrac{x}{2} = \angle PRQ = \angle SPR + PSR = 18°$, therefore $x = 36$, the answer is (A).

Example 2. Let $\triangle ABC$ be an equilateral triangle inscribed in a circle, and let M be a point on the arc BC. Prove that $MA = MB + MC$.

Solution   It is clear that $AM > CM$. On AM take a point D such that AD

$= CM$. It suffices to show that $MB = MD$. Since $AB = BC, AD = CM$ and $\angle BAM = \angle BCM$,

$$\therefore \triangle ABD \cong \triangle CBM.$$

Thus, $BM = BD$ and $\angle ABD = \angle CBM$. For the $\triangle BMD$, $\angle DBM = \angle ABC = 60°$, Thus, $\triangle DBM$ is equilateral, so $MB = MD$.

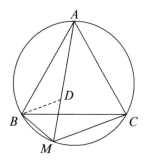

Example 3. (SSSMO/2009) Let $ABCD$ be a quadrilateral inscribed in a circle with diameter AC, and let E be the foot of perpendicular from D onto AB, as shown in the figure below. If $AD = DC$ and the area of quadrilateral $ABCD$ is 24 cm^2, find the length of DE in cm.

    (A) $3\sqrt{2}$;   (B) $2\sqrt{6}$;   (C) $2\sqrt{7}$;   (D) $4\sqrt{2}$;   (E) 6.

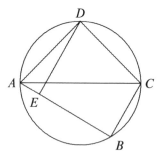

Solution   Let r be the radius of the circle and $\angle CAB = \theta$. Connect DB. Then

$$[ABCD] = [DAB] + [DCB] = \frac{1}{2}AD \cdot AB \sin \angle DAB + \frac{1}{2}CD \cdot CB \sin \angle BCD.$$

$$24 = \frac{1}{2}\sqrt{2}r \cdot 2r \sin(90° - \theta) \cdot \sin(45° + \theta)$$

$$+ \frac{1}{2}\sqrt{2}r \cdot 2r \sin\theta \cdot \sin(135° - \theta)$$

$$= r^2[\cos\theta(\sin\theta + \cos\theta) + \sin\theta(\sin\theta + \cos\theta)]$$

$$= r^2(\sin\theta + \cos\theta)^2 = [\sqrt{2}r \cdot \sin(45° + \theta)]^2$$

$$= DE^2.$$

$$\therefore DE = \sqrt{24} = 2\sqrt{6}, \text{ the answer is (B)}.$$

Example 4. (CHINA/2002) In the circle $\odot O$, the radius $r = 5$ cm, AB and CD are two parallel chords and $AB = 8$ cm, $CD = 6$ cm. Find the length of the chord AC.

Solution   As shown in the diagrams (1) and (2) below, there are four possible cases.

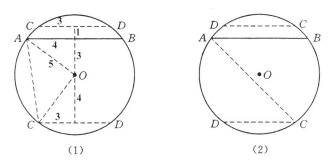

(1)   (2)

In the diagram (1), the longer length of AC is given by

$$AC = \sqrt{(4-3)^2 + (4+3)^2} = \sqrt{50} = 5\sqrt{2};$$

and the shorter length of AC is given by

$$AC = \sqrt{(4-3)^2 + 1^2} = \sqrt{2}.$$

Similarly, as shown in the diagram (2), the longer length of AC is given by

$$AC = \sqrt{(4+3)^2 + (4+3)^2} = 7\sqrt{2};$$

and the shorter length of AC is given by

$$AC = \sqrt{(4+3)^2 + 1^2} = 5\sqrt{2}.$$

Thus, the length of the chord AC may be $\sqrt{2}, 5\sqrt{2}$ or $7\sqrt{2}$.

Example 5. (BMO/2003) The triangle ABC, where $AB < AC$, has a circumcircle S. The perpendicular from A to BC meets S again at P. The point X lies on the line segment AC, and BX meets S again at Q.

Show that $BX = CX$ if and only if PQ is a diameter of S.

Solution   As shown in the diagram,

$$\angle CBP + \angle BCA = \angle CAP + \angle BCA = 90°.$$

$$\angle CBP + \angle QBC = \angle QBP.$$

$$\therefore PQ \text{ is a diameter of } S \Leftrightarrow \angle QBP = 90°$$

$$\Leftrightarrow \angle CBP + \angle QBC = 90°$$

$$\Leftrightarrow \angle BCA = \angle QBC$$

$$\Leftrightarrow BX = CX.$$

Example 6. (CHINA/2005) Given that the $\triangle ABC$ is inscribed in the circle $\odot O$, such that the diameter CD is perpendicular to AB at E. The chord BF intersects CD and AC at M and N respectively, and $BF = AC$. Connect AD, AM. Prove that

(i) $\triangle ACM \cong \triangle BCM$;

(ii) $AD \cdot BE = DE \cdot BC$;

(iii) $BM^2 = MN \cdot MF$.

Solution   (i)   Since CD is the perpendicular bisector of the chord AB, so

$$AM = BM, AC = BC,$$

and CM is shared, therefore $\triangle ACM \cong$ $\triangle BCM$ (S.S.S.).

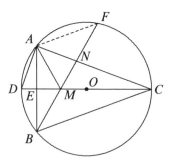

(ii) $AD \cdot BE = DE \cdot BC \Leftrightarrow \dfrac{BE}{DE} = \dfrac{BC}{AD}$.
From

$$\angle ADE = \angle ADC = \angle CBA = \angle CBE$$

and

$$\angle DAE = \angle DAB = \angle D = \angle BCE,$$

it follows that $\triangle DAE \sim \triangle BCE$, so $\dfrac{BE}{DE} = \dfrac{BC}{AD}$.

(iii) Connect AF. $BF = AC \Rightarrow \overset{\frown}{BAF} = \overset{\frown}{AFC}$, so

$$\overset{\frown}{AB} = \overset{\frown}{FC}$$

which implies that $\angle CBF = \angle AFB$. Since $\triangle ACM \cong \triangle BCM$,

$$\angle AFM = \angle AFB = \angle CBF = \angle CBM = \angle CAM = \angle NAM,$$

therefore $\triangle MAN \sim \triangle MFA$, so $\dfrac{MN}{MA} = \dfrac{MA}{MF}$, namely

$$MA^2 = MN \cdot MF, \quad \text{or} \quad BM^2 = MN \cdot MF$$

since $BM = AM$.

Example 7. (CHNMOL/2004) The circumcircle $\odot O$ of the quadrilateral $ABCD$ has a radius 2. AC and BD intersect at E, such that $AE = EC$. Given that $AB = \sqrt{2}AE$, $BD = 2\sqrt{3}$, find the area of $ABCD$.

Solution $AE = EC$ and $AB = \sqrt{2}AE$ give $AB^2 = 2AE^2 = AE \cdot AC$, so $\dfrac{AB}{AC} = \dfrac{AE}{AB}$. Since $\angle EAB = \angle BAC$, it follows that

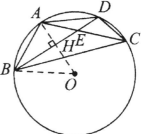

$\triangle ABE \sim \triangle ACB$, so that $\angle ABE = \angle ACB = \angle ADB$, $\therefore AB = AD$.

Therefore A bisects the arc BAD, so $OA \perp BD$. Let H be the point of intersection of AO and BD, then $OH \perp BH$ and

$$OH = \sqrt{2^2 - (\sqrt{3})^2} = 1, AH = OA - OH = 1 \Rightarrow [ABD] = \frac{1}{2} \cdot 2\sqrt{3} \cdot 1 = \sqrt{3}.$$

Since $AE = CE \Rightarrow [BCE] = [ABE], [CDE] = [ADE] \Rightarrow [CBD] = [ABD]$, it is obtained that

$$[ABCD] = 2[ABD] = 2\sqrt{3}.$$

Example 8. Given that P is a fixed point on the angle bisector of the $\angle CAB$. Construct a circle passing through A and P. If the circle intersects the sides AB and AC at points M and N respectively, prove that the value of $AM + AN$ is independent of the choice of the circle.

Solution For any chosen circle that passes through A and P, let $d = AP$ and write $\angle MAP = \angle NAP = \alpha$. Connect MP, NP.

Applying the cosine rule to $\triangle MAP$ and $\triangle PAN$ respectively, since $PM = PN$, it follows that

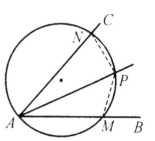

$$PM^2 = d^2 + AM^2 - 2d \cdot AM \cos\alpha,$$
$$PN^2 = d^2 + AN^2 - 2d \cdot AN \cos\alpha,$$

or

$$AM^2 - 2d \cos\alpha \cdot AM + (d^2 - PM^2) = 0$$
$$AN^2 - 2d \cos\alpha \cdot AN + (d^2 - PM^2) = 0.$$

The values of PM, PN are given if the circle is drawn, then the corresponding values of AM and AN are the roots of the quadratic equation

$$x^2 - 2d \cos\alpha \cdot x + (d^2 - PM^2) = 0,$$

and by the Viete's theorem, $AM + AN = 2d \cos\alpha$, which is independent of the choice of the circle.

Testing Questions   (A)

1. (SSSMO/2010) In the figure below, AB and CD are parallel chords of a circle with centre O and radius r cm. It is given that $AB = 46$ cm, $CD = 18$ cm and $\angle AOB = 3\angle COD$. Find the value of r.

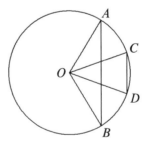

2. (CHNMOL/2004) If the shortest distance of a given point to a given circle is 4 cm and the longest distance is 9 cm, then the radius of the circle is

    (A) 2.5 cm;   (B) 2.5 cm or 6.5 cm;   (C) 6.5 cm;   (D) 5 cm or 13 cm.

3. (IRE/2003) P, Q, R, S are four distinct points on the circle $\odot O$, where PS is the diameter, $QR \parallel PS$, and PR, QS intersect at A. B is a point in the plane such that the quadrilateral $POAB$ is a parallelogram. Prove that $BQ = BP$.

4. (SMO/2003) Let ABC be an equilateral triangle inscribed in a circle and P a point on the minor arc BC. Suppose that AP intersects BC at D with $PB = 21$ and $PC = 28$. Find PD.

5. (CHINA/2004) $\odot O$ is a circle of radius 1, the point P is on the circle. If A and B are on $\odot O$ such that $\angle APB = \angle AOB$, find the length of the chord AB.

6. (CMO/2004) Let A, B, C, D be four points on a circle (occurring in clockwise order), with $AB < AD$ and $BC > CD$. Let the bisector of angle BAD meet the circle at X and the bisector of angle BCD meet the circle at Y. Consider the hexagon formed by these six points on the circle. If four of the six sides of the hexagon have equal length, prove that BD must be a diameter of the circle.

7. (CHNMO/TST/2003) A convex n-sided polygon has a circumcircle and an inscribed circle, its area is B, and the areas of its circumcircle circle and inscribed circle are A and C respectively. Prove that $2B < A + C$.

8. (GERMANY/2005) Let A, B, C be three distinct points on the circle $\odot O$. By passing through B, C respectively make lines h and g such that $h \perp BC$ at B and $g \perp BC$ at C. Given that the perpendicular bisector of AB intersects h at F, and the perpendicular bisector of AC intersects g at G. Prove that $BF \cdot CG$ is independent of the choice of A when fixing B and C.

9. (INDIA/2006) In a cyclic quadrilateral $ABCD, AB = a, BC = b, CD = c, \angle ABC = 120°, \angle ABD = 30°$. Prove that

   (i) $c \geq a + b$;

   (ii) $|\sqrt{c+a} - \sqrt{c+b}| = \sqrt{c-a-b}$.

10. (BMO/2005) Let ABC be an acute-angled triangle, and let D, E be the feet of the perpendiculars from A, B to BC, CA respectively. Let P be the point where the line AD meets the semicircle constructed outwardly on BC, and Q be the point where the line BE meets the semicircle constructed outwardly on AC. Prove that $CP = CQ$.

Testing Questions   (B)

1. (BMO/2003) Let ABC be a triangle and let D be a point on AB such that $4AD = AB$. The half-line ℓ is drawn on the same side of AB as C,

starting from D and making an angle of θ with DA where $\theta = \angle ACB$. If the circumcircle of ABC meets the half-line ℓ at P, show that $PB = 2PD$.

2. (CHNMOL/2005) It is given that the $\triangle ABC$ is inscribed in the $\odot O$, AD is the diameter of the $\odot O$. The points E and F are on the extensions of AB and AC respectively, and the line segment EF intersects the $\odot O$ and AD at M, N and H respectively, where H is the midpoint of OD.

    Given that $MD = DN$, $EH - HF = 2$, $\tan \alpha = \dfrac{3}{4}$, where $\angle ACB = \alpha$,

    and EH, HF are two real roots of the equation $x^2 - (k + 2)x + 4k = 0$.

    (i)   Find EH and HF.   (ii)   Find BC.

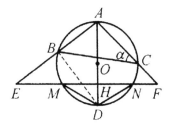

3. (BELARUS/2003) Given that $ABCD$ is a convex quadrilateral, and $AC \perp BD$ at O. Suppose that O_1, O_2, O_3 and O_4 are the centers of the inscribed circles of $\triangle AOB$, $\triangle BOC$, $\triangle COD$ and $\triangle DOA$ respectively. Prove that

    (i)   the sum of diameters of the $\odot O_1, \odot O_2, \odot O_3$ and $\odot O_4$ is not greater than $(2 - \sqrt{2})(AC + BD)$.

    (ii) $O_1 O_2 + O_2 O_3 + O_3 O_4 + O_4 O_1 < 2(\sqrt{2} - 1)(AC + BD)$.

4. (AUSTRIA/2005) Construct the semi-circle Γ with the diameter AB and the midpoint M. Now construct the semi-circle Γ_1 with the diameter MB on the same side as Γ. Let X and Y be points on Γ_1, such that the arc $\overset{\frown}{BX}$ is 1.5 times of the arc $\overset{\frown}{BY}$. The line MY intersects the line BX in D and the semi-circle Γ in C. Show that Y is the midpoint of CD.

5. (RUSMO/2009) Let ABC be a given triangle and BD ($D \in AC$) its interior angle bisector. The line BD intersects the circumcircle Ω of triangle ABC at B and E. Circle ω with diameter DE cuts Ω again at F. Prove that BF is the symmetric line of a median of the triangle ABC with respect to the line BD.

Lecture 11

Relation of Line and Circle and Relation of Circles

The main focus for investigating relation between line and circle is their tangency.

(i)   A line l is tangent to a circle $\odot O$ if and only if l meets $\odot O$ at one point A such that $l \perp OA$. In this case A is call the *tangent point* or the *point of contact*, and l is called *tangent to the circle at A*.

(ii)   When the two tangent line introduced from any exterior point P of a circle $\odot O$ have two tangent points A and B, then

$$PA = PB, \quad \angle PAB = \angle PBA; \quad\quad OP \perp AB, \quad \angle APO = \angle BPO.$$

(iii)   *Alternate segment theorem*: An angle included by a tangent and a chord through the tangent point is equal to the angle in the alternate segment.

(iv)   *Intersecting chords theorem*: When two chords AB and CD intersect at an interior point P of a circle, then $AP \cdot PB = CP \cdot PD$.

(v)   The circle which is tangent to three sides of a triangle is called the inscribed circle or shortly *incircle* of the triangle. Its center, called *incenter*, is the point of intersection of three angle bisectors of the three interior angles, and its radius is called *inradius*.

The analysis of relation among circles is based on that of two circles.

(vi)   When two circles $\odot O_1$ and $\odot O_2$ are intersected at points A and B, then the line $O_1 O_2$ is the perpendicular bisector of AB, there are two *external common tangents*, and they are symmetric with respect to the line $O_1 O_2$.

(vii)   When two circles $\odot O_1$ and $\odot O_2$ are tangent (internally or externally), there is one *common tangent* which is perpendicular to $O_1 O_2$. When $\odot O_1$ and $\odot O_2$ are externally tangent each other, there are two external common tangents also, which are symmetric with respect to the line $O_1 O_2$.

(viii)   When two circles $\odot O_1$ and $\odot O_2$ are separated externally, they have a pair of external common tangents and a pair of *internal common tangents*, and each pair is symmetric with respect to the line $O_1 O_2$.

Examples

Example 1. Prove that a quadrilateral has an inscribed circle if and only if its two sums of the lengths of two opposite sides are equal, in particular, a parallelogram has an inscribed circle if and only if it a rhombus.

Solution   As shown in the figure below, Suppose that the quadrilateral has an inscribed circle, and E, F, G, H are the tangent points of the circle on the sides DA, AB, BC and CD respectively. Then $AE = AF, BF = BG, CG = CH$ and $DH = DE$ implies that

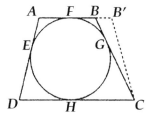

$$
\begin{aligned}
AB + CD &= AF + BF + CH + DH \\
&= AE + BG + CG + DE \\
&= (AE + DE) + (BG + CG) \\
&= AD + BC.
\end{aligned}
$$

Conversely, if suppose that the quadrilateral $ABCD$ satisfies the condition $AB + CD = BC + AD$, then it is always possible to make a circle such that it touches the sides AB, CD and AD at some points E, F, H on AD, AB, CD respectively.

If CB is not tangent to the circle, then make the second tangent line from C to the circle. Suppose that the tangent intersects the line AB at some point B', where $B \neq B'$. From above reasoning it follows that

$$AB' + CD = B'C + AD.$$

Then

$$AB + CD = BC + AD \Rightarrow |AB - AB'| = |BC - B'C| \Rightarrow BB' = |BC - B'C|.$$

However, applying the triangle inequality to the $\triangle BB'C$ shows it is impossible. Thus, B and B' must coincide, i.e., CB must be a tangent to the circle, namely, the circle is the inscribed circle of $ABCD$.

When a parallelogram $ABCD$ has an inscribed circle, then $AB = CD, BC = AD$ and $2AB = AB + CD = BC + AD = 2BC$, so $AB = BC = CD = AD$, i.e., $ABCD$ is a rhombus. Conversely, if $ABCD$ is a rhombus, the conclusion then is clear from above reasoning.

Example 2. (CMC/2009) Given that PA, PB are the tangent lines from P to the circle $\odot O$, and the line segment PCD is a transversal of $\odot O$, and E is the point of intersection of AB and PD. Prove that $\dfrac{PC}{PD} = \dfrac{CE}{DE}$.

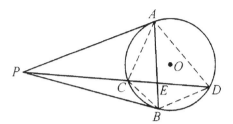

Solution Connect AC, AD, BC and BD. Then $\dfrac{PC}{PD} = \dfrac{[PAC]}{[PAD]} = \dfrac{[PBC]}{[PBD]}$.
By the alternate segment theorem,

$$\angle PAC = \angle PDA \Rightarrow \triangle PAC \sim \triangle PDA$$

and

$$\angle PBC = \angle PDB \Rightarrow \triangle PBC \sim \triangle PDB,$$

therefore $\dfrac{[PAC]}{[PAD]} = \dfrac{AC^2}{AD^2}, \dfrac{[PBC]}{[PBD]} = \dfrac{BC^2}{BD^2}$. Thus, $\dfrac{AC}{AD} = \dfrac{BC}{BD}$ and

$$\frac{PC}{PD} = \frac{AC^2}{AD^2} = \frac{AC}{AD} \cdot \frac{BC}{BD}. \tag{11.1}$$

Since $\triangle ACE \sim \triangle DBE$ and $\triangle BCE \sim \triangle DAE$,

$$\frac{AC}{DB} = \frac{AE}{DE} \quad \text{and} \quad \frac{BC}{DA} = \frac{CE}{AE}. \tag{11.2}$$

Thus, the combination of (11.1) and (11.2) gives

$$\frac{PC}{PD} = \frac{AC}{AD} \cdot \frac{BC}{BD} = \frac{AC}{DB} \cdot \frac{BC}{DA} = \frac{AE}{DE} \cdot \frac{CE}{AE} = \frac{CE}{DE},$$

as desired.

Example 3. (CMC/2009) It is given that the inradius of the $\triangle ABC$ is 2, and $\tan A = -\dfrac{4}{3}$, find the minimum value of the area of $\triangle ABC$.

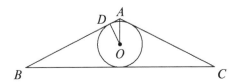

Solution Let $BC = a, CA = b, AB = c, OD \perp AB$ at D, then

$$b + c - a = 2AD \Rightarrow AD = \frac{b + c - a}{2}.$$

Let $x = \tan \dfrac{A}{2}$. Since $-\dfrac{4}{3} = \dfrac{2x}{1 - x^2} \Rightarrow 2x^2 - 3x - 2 = 0$, then

$$2x^2 - 3x - 2 = 0 \Leftrightarrow (2x + 1)(x - 2) = 0 \Rightarrow x = 2 \ (\because x > 0),$$

therefore $2 = DO = \tan \dfrac{A}{2} \cdot AD = 2AD \Rightarrow AD = 1 \Rightarrow b + c - a = 2$. Since

$$\tan A = -\frac{4}{3} \Rightarrow \sin A = \frac{4}{5}, \quad \cos A = -\frac{3}{5},$$

so $a + b + c = \dfrac{a + b + c}{2} \cdot 2 = [ABC] = \dfrac{1}{2} bc \sin A = \dfrac{2}{5} bc$. From

$$a + b + c = 2(b + c) - (b + c - a) = 2(b + c) - 2$$

it follows that $bc = 5(b + c) - 5 \geq 10\sqrt{bc} - 5$, i.e., $(\sqrt{bc})^2 - 10\sqrt{bc} + 5 \geq 0$. Therefore

$$\sqrt{bc} \geq \frac{10 + \sqrt{80}}{2} = 5 + 2\sqrt{5}, \quad \text{or} \quad bc \geq 45 + 20\sqrt{5}.$$

Thus, $[ABC] = \dfrac{2}{5} bc \geq 18 + 8\sqrt{5}$, and the equality holds when $b = c = 5 + 2\sqrt{5}$.

Example 4.

(CHINA/2005) AB is the diameter of the $\odot O$ with $AB = a$. C is a point on the tangent line to the circle at A, and $AC = AB$.
The line segment OC intersects $\odot O$ at D, and the extension of BD intersects AC at E, as shown in the given diagram.
Find the length of AE.

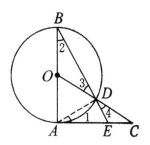

Solution Connect AD. Then AD is the altitude of the $\text{Rt}\triangle ABE$ on BE. From $\triangle ADE \sim \triangle BDA$,

$$\frac{AE}{DE} = \frac{AB}{AD} = \frac{AC}{AD}.$$

By the alternate segment theorem, $\angle 1 = \angle 2$. $OB = OD$ implies $\angle 2 = \angle 3$, so $\angle 1 = \angle 2 = \angle 3 = \angle 4$, therefore $\triangle CDE \sim \triangle CAD$. Then

$$\frac{AC}{AD} = \frac{CD}{DE}.$$

Therefore $\dfrac{AE}{DE} = \dfrac{CD}{DE}$, namely $AE = CD$.

$\triangle CDE \sim \triangle CAD$ implies also $\dfrac{CD}{AC} = \dfrac{CE}{CD}$, i.e., $CD^2 = CE \cdot AC$, therefore

$$AE^2 = CE \cdot AC.$$

Let $AE = x$, then $CE = a - x$ and $x^2 = a(a-x)$ which gives $x^2 + ax - a^2 = 0$, thus

$$AE = x = \frac{-a + \sqrt{a^2 + 4a^2}}{2} = \frac{\sqrt{5} - 1}{2} a \text{ (the negative root is not acceptable).}$$

Example 5.

(CMC/2010) $\triangle ABC$ is inscribed in the $\odot O$ with $AB = AC$. The line MN is tangent to the circle at C, $BD \parallel MN$ and AC intersects BD at E, as shown in the given diagram.
(i) Prove that $\triangle ABE \cong \triangle ACD$;
(ii) Find AE if $AB = 6$, $BC = 4$.

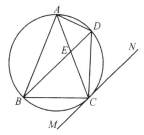

Solution   (i)   For triangles ABE and ACD, since $AB = AC, \angle ABE = \angle ACD$ and by the alternate segment theorem and $BD \parallel MN$,

$$\angle CAD = \angle NCD = \angle BDC = \angle BAC = \angle BAE,$$

therefore $\triangle ABE \cong \triangle ACD$ (S.A.A.).
(ii)   Then

$$\angle BAC = \angle CAD, \; BE = CD \;\; \Rightarrow \;\; \angle BDC = \angle CBD, \; BE = CD$$
$$\Rightarrow \;\; BE = CD = BC = 4.$$

Let $AE = x$. Then $\triangle AEB \sim \triangle DEC$ implies that

$$DE = AE \cdot \frac{CD}{AB} = \frac{4x}{6} = \frac{2x}{3}.$$

The intersecting chords theorem yields $x(6 - x) = 4 \cdot \dfrac{2x}{3}$, then it gives that

$$x = 6 - \frac{8}{3} = \frac{10}{3}. \qquad \text{Thus, } AE = \frac{10}{3}.$$

Example 6. (CMC/2010) A circle of radius 2 and a circle of radius 3 are tangent externally at T. The line MN is an external common tangent, where M, N are the two tangent points. Find the value of $\dfrac{MT}{NT}$.

Solution   As shown in the right diagram, let the radii of the $\odot C$ and $\odot D$ be 2 and 3 respectively, and let $\angle MCT = \theta$, then $\angle NDT = \pi - \theta$. By the alternate segment theorem,

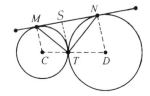

$$\angle NMT = \frac{1}{2}\theta, \angle MNT = \frac{1}{2}(\pi - \theta).$$

Therefore $\angle MTN = \dfrac{\pi}{2}$. From T introduce $TS \perp MN$ at S, Then TS is the altitude on the hypotenuse of the $\mathrm{Rt}\triangle MTN$. By the projection theorem of right triangles,

$$\frac{MT^2}{NT^2} = \frac{MS \cdot MN}{NS \cdot NM} = \frac{MS}{NS} = \frac{CT}{DT} = \frac{2}{3} \Rightarrow \frac{MT}{NT} = \sqrt{\frac{2}{3}} = \frac{\sqrt{6}}{3}.$$

Example 7. (MACAO/2001) A big circle $\odot O$ of radius R, a circle $\odot B$ of radius $2r$, and two circles $\odot A, \odot A'$ of radius r are tangent pairwise, as shown in the following left diagram. Find the ratio $\dfrac{r}{R}$.

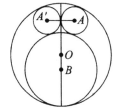
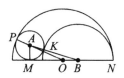

Solution   By symmetry, the centers O and B are on the internal common tangent of the $\odot A$ and $\odot A'$, so it suffices to consider a half of the graph, as shown in the right digram above.

Let $MO = x, MB = y$, then $MN = x + R = y + 2r$, so $x = y + 2r - R$. Since $AO = R - r, AB = 3r$, by the Pythagoras' theorem,

$$y^2 = MB^2 = AB^2 - AM^2 = 8r^2 \Rightarrow y = \sqrt{8}r, x = (2 + \sqrt{8})r - R.$$
$$x^2 = MO^2 = AO^2 - AM^2 = (R - r)^2 - r^2 = R^2 - 2Rr,$$

so $[(2 + \sqrt{8})r - R]^2 = R^2 - 2Rr$, which gives $(2 + 2\sqrt{8})R = (2 + \sqrt{8})^2 r$, thus,

$$\frac{r}{R} = \frac{2 + 2\sqrt{8}}{(2 + \sqrt{8})^2} = \frac{1 + 2\sqrt{2}}{2(\sqrt{2} + 1)^2}$$
$$= \frac{(1 + 2\sqrt{2})(\sqrt{2} - 1)^2}{2} = \frac{4\sqrt{2} - 5}{2}.$$

Testing Questions (A)

1. (CHINA/2005) $ABCD$ is an inscribed square of the circle $\odot O$. P is the midpoint of the minor arc $\overarc{AB}$, PD intersects AB at E. Find the ratio $\dfrac{PE}{DE}$.

2. (SSSMO/2010) In the figure below, ABC is an isosceles triangle inscribed in a circle with centre O and diameter AD, with $AB = AC$. AD intersects BC at E, and F is the midpoint of OE. Given that BD is parallel to FC and $BC = 2\sqrt{5}$ cm, find the length of CD in cm.

   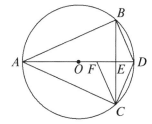

   (A) $\dfrac{3\sqrt{5}}{2}$; (B) $\sqrt{6}$; (C) $2\sqrt{3}$;
   (D) $\sqrt{7}$; (E) $2\sqrt{6}$.

3. (FINLAND/2004) Two circles of radii r and R respectively are tangent externally. Find the length of the line segment on the internal common tangent which is between the two external common tangent lines of the two circles.

4. (NEW ZEALAND/2004) Given that four circles are inside a convex quadrilateral, such that each are tangent to two sides of the quadrilateral, and each are tangent to other two circles externally. If the quadrilateral has an inscribed circle, prove that at least two of them have equal radii.

5. (THAILAND/2004) It is given that the circle ω is the circumcircle of equilateral $\triangle ABC$, and the circle ω_1 is tangent to ω externally at a point differing from A, B, C. The points A_1, B_1, C_1 are on ω_1 such that AA_1, BB_1 and CC_1 are tangent to ω_1. Prove that the length of one of AA_1, BB_1, CC_1 is equal to the sum of lengths of the other two.

6. (CMC/2009) When put some smaller circles of radii 1 into a big circle of radius 11, such that each small circle is tangent to the big circle internally, and any two small circles have no overlapped part. How may small circles can be put in at most?

   (A) 30; (B) 31; (C) 32; (D) 33.

7. (JAPAN/2008) $ABCD$ is a square of side 1, M is center of the circle taking AD as the diameter, E is a point on the side AB such that CE is tangent to $\odot M$. Find the area of $\triangle CBE$.

8. (SMO/2009) Let O be the center of the circle inscribed in a rhombus $ABCD$. Point E, F, G, H are chosen on sides AB, BC, CD and DA respectively so that EF and GH are tangent to the inscribed circle. Show that EH and FG are parallel.

9. (**Newton's Theorem**) Given that the convex quadrilateral $ABCD$ has an inscribed circle, and the circle touches the sides AB, BC, CD, DA at E, F, G, H respectively. Prove that the line segments AC, BD, EG, FH intersect at one common point.

Testing Questions   (B)

1. (CMC/2008) It is given that the circle $\odot O$ touches the sides AB and AC of the $\triangle ABC$ at P and Q respectively, and tangent internally to the circumcircle of $\triangle ABC$ at D, as shown in the left diagram below. Prove that $\angle POQ = 2\angle MDC$, where M is the midpoint of PQ.

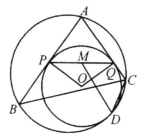 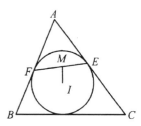

2. (CMC/2010) Given that the incenter of $\triangle ABC$ is I, and the $\odot I$ touches AC, AB at E and F respectively. M is a point on the line segment EF, as shown in above right diagram. Prove that the areas of triangles MAB and MAC are equal if and only if $MI \perp BC$.

3. (AIME/2010) In $\triangle ABC$ with $AB = 12$, $BC = 13$, and $AC = 15$, let M be a point on AC such that the incircles of $\triangle ABM$ and $\triangle BCM$ have equal radii. Let p and q be positive relatively prime integers such that $\dfrac{AM}{CM} = \dfrac{p}{q}$. Find $p + q$.

4. (ESTONIA/TST/2008) A, B are two fixed points on the circle Γ_1. The circle Γ_2 is tangent to AB at B and its center is on Γ_1. The line ADE intersects Γ_2 at D, E, and the line BD intersects Γ_1 again at F. Prove that BE is tangent to Γ_1 if and only if $DF = DB$.

Lecture 12

Cyclic Polygons

An n-sided polygon ($n \geq 4$) is said to be a **cyclic polygon** if it is inscribed in a circle. In this case, the n vertices are said to be **concyclic**.

The main part of this lecture is on cyclic quadrilateral, since it is foundation for investigating cyclic polygon.

Criteria for determining a cyclic quadrilateral

(I)   A quadrilateral $ABCD$ is cyclic if and only if its four interior angles satisfy $\angle A + \angle C = \angle B + \angle D = 180°$.

(II)   A convex quadrilateral is cyclic if and only if any exterior angle equals to its opposite interior angle.

(III)   A convex quadrilateral $ABCD$ is cyclic if and only if the two angles subtended by a same side of $ABCD$ are equal, for example, $\angle ACB = \angle ADB$, which are subtended by the side AB.

(IV)   (*Inverse intersecting chords theorem*) When two line segments AC and BD are intersected at a point P, the quadrilateral $ABCD$ is cyclic if if $AP \cdot PC = BP \cdot BD$.

(V)   When two line segments AC and BD are not intersected but their extensions are intersected at a point P, then the quadrilateral $ABCD$ is cyclic if and only if $PA \cdot PD = PB \cdot PC$.

The proofs of necessity of (I) to (III) are the direct applications of the fundamental properties. The sufficiency of (I) to (V) can be proven by contradiction. The proof of necessity of (V) can be found in next lecture.

In the examples below, some are to prove that points are concyclic, and some are the applications of concyclic points for solving other geometric problems involving circles.

Examples

Example 1. (RUSMO/2008/R4) In $\triangle ABC$ $AB > AC$, the tangent line at B to the circumcircle intersects the line AC at P. D is the symmetric point of B with respect to P, E is the symmetric point of C with respect to the line BP. Prove that the quadrilateral $ABED$ is cyclic.

Solution   By the alternate segment theorem, $\angle PBC = \angle BAC = \angle PAB$, therefore $\triangle PBC \sim \triangle PAB$, so $\dfrac{PA}{PB} = \dfrac{PB}{PC}$,

$$\therefore PA \cdot PC = PB^2 = PD^2,$$

or $\dfrac{PA}{PD} = \dfrac{PD}{PC}$. Since $\angle APD$ is shared,

$$\triangle APD \sim \triangle DPC \text{ (S.A.S.)},$$
$$\therefore \angle CAD = \angle PAD = \angle PDC.$$

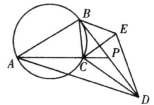

C and E are symmetric in the line BP implies that $\angle PDC = \angle PDE = \angle BDE$ and $\angle PBC = \angle DBE$, so that

$$\begin{aligned}\angle BAD &= \angle PAD + \angle PAB = \angle PDC + \angle PBC = \angle BDE + \angle DBE \\ &= 180° - \angle BED.\end{aligned}$$

Thus, the quadrilateral $ABED$ is cyclic.

Example 2. (CMC/2010) As shown in the given diagram below, $ABCD$ is a convex quadrilateral, $\angle ABC = \angle ADC$, E, F, G, H are the midpoints of AC, BD, AD, CD respectively. Prove that

    (i) E, F, G, H are concyclic;

    (ii) $\angle AEF = \angle ACB - \angle ACD$.

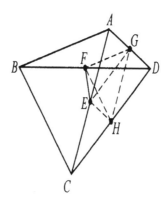

Solution (i) Connect EG, EH, FG, FH, GH, then $FG \parallel BA, FH \parallel BC$, so

$$\angle GFH = \angle ABC.$$

Similarly, $\angle GHF = \angle ACB$. Since $DGEH$ is a parallelogram, so

$$\angle GEH = \angle ADC = \angle ABC = \angle GFH,$$

therefore $EFGH$ is cyclic, i.e., E, F, G, H are concyclic.

(ii) $EFGH$ is cyclic $\Rightarrow \angle GEF = \angle GHF = \angle ACB$, and $EG \parallel CD$ implies that $\angle AEG = \angle ACD$, therefore

$$\angle AEF = \angle GEF - \angle AEG = \angle ACB - \angle ACD.$$

Example 3. (SNOVENIA/TST/2008-2009) In the acute triangle ABC, D is on AB, the circumcircles of $\triangle BCD$ and $\triangle ADC$ intersect AC and BC at the points E, F respectively. Let the circumcenter of $\triangle CEF$ be O. Prove that the circumcenters of $\triangle ADE, \triangle ADC, \triangle DBF, \triangle DBC$ and the points D and O are concyclic, and $OD \perp AB$.

Solution Let $O_1, O_2, O_3,$ O_4 be the circumcenters of triangles ADE, ADC, DBF, DBC respectively. Then $O_1 O_2$ is on the perpendicular bisector of AD, and $O_3 O_4$ is on the perpendicular bisector of DB. Use α, β, γ to denote the interior angles of $\triangle ABC$ and let T_1, T_2, T_3, T_4, T_5 be the midpoints of the line segments AD, CF, BD, CE, CD respectively.

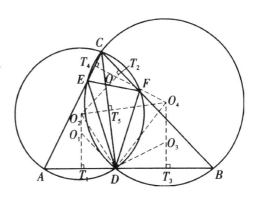

$ADFC$ is cyclic gives $\angle DFB = \alpha$. O_3 is the circumcenter of $\triangle DBF$ implies that $\angle DO_3 T_3 = \alpha$, and similarly, $\angle DO_2 T_5 = \angle DAC = \alpha$.

Besides, O_2, T_5, O_4 are collinear, so $\angle DO_2 O_4 = \angle DO_2 T_5 = \alpha = \angle DO_3 T_3$, therefore

O_2, O_4, O_3, D are concyclic. Similarly, O_4, D, O_1, O_2 are concyclic.

Therefore O_1 and O_3 are both on the circumcircle of $\triangle O_2 O_4 D$. Besides, $\angle DO_2 O_4 = \alpha, \angle O_2 O_4 D = \beta$ implies that $\angle O_4 D O_2 = \gamma$.

On the other hand, $CT_4 OT_2$ is cyclic, so $\angle T_4 OT_2 = 180° - \gamma$, therefore O, O_2, D, O_4 are concyclic. Thus, O_1, O_2, O_3, O_4, O, D are concyclic.

Since $\angle OO_4O_3 = \angle T_4O_4T_3 = 180° - \angle CAB = 180° - \alpha$, and $\angle DOO_4 = \angle DO_3T_3 = \alpha$, so $OD \parallel O_3O_4$, hence $OD \perp AB$.

Example 4. (CROATIA/2009) $ABCD$ is a convex quadrilateral, the circumcircle of $\triangle ABC$ intersects CD, DA at P, Q respectively, and the circumcircle of $\triangle ACD$ intersects AB, BC at R, S respectively. The lines BP, BQ intersect the line RS at the points M, N respectively. Prove that M, N, Q, P are concyclic.

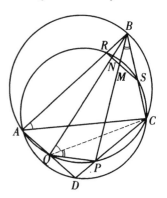

Solution   The angles at circle subtended by a same chord are equal, so
$\angle BQC = \angle BAC, \angle CQP = \angle CBP$. Since $\angle BSR = \angle RAC = \angle BAC = \angle BQC$, hence

$$\angle BMN = 180° - \angle BMS$$
$$= \angle SBM + \angle BSM = \angle CBP + BSR$$
$$= \angle CQP + \angle BQC = \angle BQP = \angle NQP.$$

Thus, M, N, Q, P are concyclic.

Example 5. (ITALY/TST/2009) Given that $\odot O_1$ and $\odot O_2$ intersect at M and N. the common tangent which is closer to M is tangent to the circles at A and B respectively. C and D are the symmetric points of A and B with respect to M respectively. The circumcircle of $\triangle DCM$ intersects $\odot O_1$ and $\odot O_2$ at E and F (which defer from M) respectively. Prove that the circumradii of $\triangle MEF$ and $\triangle NEF$ are equal.

Solution   Take the point N' such that $NEN'F$ is a parallelogram.

Suppose that the Extensions of AD and BC intersect $\odot O_1$ and $\odot O_2$ at E' and F' respectively. Since M, C, F, E, D are concyclic,

$$\angle MFC = \angle MDC = \angle MBA = \angle MFB.$$

There are B, C, F are collinear, $F = F'$. Similarly, $E = E'$.
Let the ray NM intersect AB at L. Then

$$\triangle LAM \sim \triangle LNA \Rightarrow LA^2 = LM \cdot LN \text{ and}$$
similarly $LB^2 = LM \cdot LN$, therefore $LA = LB$.

AC, BD bisect each other $\Rightarrow ABCD$ is a parallelogram $\Rightarrow MN \parallel BF \parallel AE$. Therefore O_1O_2 is perpendicular to AE and BF, and A, M, B and E, N, F are symmetric with respect to the line O_1O_2, which implies that $\angle ENF = \angle AMB$ so that

$$\angle EN'F + \angle EMF = \angle ENF + \angle EMF = \angle AMB + \angle EMF$$
$$= \angle AMB + \angle EMN + \angle FMN = \angle AMB + \angle AEM + \angle BFM$$
$$= \angle AMB + \angle BAM + \angle ABM = 180°,$$

therefore $MEN'F$ is cyclic. Therefore the circumradius of $\triangle MEF$ is that of $\triangle EN'F$. Since $ENFN'$ is a parallelogram, so $\triangle EN'F \cong \triangle ENF$, so the circumradius of $\triangle EN'F$ is equal to that of $\triangle NEF$.

Example 6. (CMC/2010) M is the center of equilateral triangle $A_1A_2A_3$, N is an arbitrary point in the plane that the triangle lies, the circle taking MN as a diameter intersects the lines MA_i at B_i, $i = 1, 2, 3$. Prove that

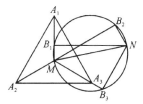

$$MB_1^2 + MB_2^2 + MB_3^2 = NB_1^2 + NB_2^2 + NB_3^2.$$

Solution The five points M, B_1, B_2, N, B_3 are concyclic, and $\angle B_1MB_3 = 120°$, so $\angle B_1B_2B_3 = 60°$ and

$$\angle B_1B_3B_2 = \angle B_1MB_2 = \angle A_1MB_2 = 60°,$$

therefore $\triangle B_1B_2B_3$ is equilateral. Let s be the length of its sides.

Lemma: If P is a point on the circumcircle of equilateral triangle ABC, then the value of

$$PA^2 + PB^2 + PC^2$$

is independent of the choice of P.

In fact, let a be the length of AB, $PA = x$, $PB = y$, $PC = z$, then $x = y + z$ (cf. Example 2 in Lecture X), therefore, by applying the cosine rule,

$$x^2 + y^2 + z^2 = (y + z)^2 + y^2 + z^2 = 2(y^2 + z^2 + yz)$$
$$= 2BC^2 = 2a^2,$$

the Lemma is proven. By the lemma,

$$MB_1^2 + MB_2^2 + MB_3^2 = NB_1^2 + NB_2^2 + NB_3^2 = 2s^2.$$

Example 7. (VIETNAM/2009) Let A, B be two fixed points and C is a variable point such that $\angle ACB = \alpha$ which is a constant and $0° < \alpha < 180°$. Let D, E, F

be the projections of the incenter I of $\triangle ABC$ on its sides BC, CA, AB respectively. Denote by M, N the intersections of the lines AI, BI with the line segment EF, respectively. Prove that the length of MN is constant and the circumcircle of $\triangle DMN$ always passes through a fixed point.

Solution   Since $\angle CEF = 90° - \frac{1}{2}\angle C = 180° -$
$\angle AIB = \angle AIN$, so I, N, E, A are concyclic.
Since I, E, A, D are concyclic also, so I, N, E, A, D
are concyclic.
Similarly, B, F, M, I, D are concyclic. Therefore

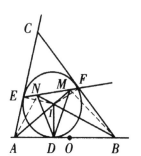

$$\angle ANB = \angle AEI = 90°, \angle AMB = \angle BFI = 90°.$$

Therefore M, N are both on the circle taking AB as
the diameter.

Taking O as the midpoint of AB,   then $\angle MON = 2\angle MAN = 2\angle MBN = \angle MAN + \angle MBN = \angle NDI + \angle MDI = \angle MDN$, so M, N, D, O are concyclic, hence the circum circle of $\triangle MND$ passes through the fixed point O.
On the other hand,

$$MN = AB \sin \angle NAM = AB \sin \angle IEF = AB \sin \frac{C}{2}$$

which is a constant.

Example 8. (BELARUS/2009) Let X, X_1, X_2 be points on the sides AB, AC, BC of $\triangle ABC$ respectively, such that $XX_1 \perp AC, X_1X_2 \perp BC, X_2X \perp AB$. Let Y, Y_1, Y_2 be points on BC, AC, AB respectively, such that $YY_1 \perp AC, Y_1Y_2 \perp AB$. Prove that $Y_2Y \perp BC$ if $XY \parallel AC$.

Solution   As shown in the graph below, when $XY \parallel AC$, then $\angle YXX_1 = \angle XX_1Y_1 = 90°$. Considering $\angle X_1X_2Y = 90°$, it follows that X, X_1, X_2, Y are concyclic, and X_1Y is a diameter of the circle. Use Γ to denote the circle.
Since $\angle YY_1X_1 = 90°$, so Y_1 is also on Γ.
Since $\angle XY_2Y_1 = 90°$ and $\angle XYY_1 = 90°$, so
X, Y_2, Y_1, Y are concyclic. Therefore Y_2 is on Γ
also.
Since $\angle X_2XY_2 = 90°$, so X_2Y_2 is the diameter of
Γ. Thus,

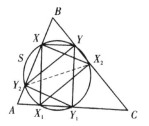

$$\angle Y_2YX_2 = 90° \Rightarrow Y_2Y \perp BC.$$

Testing Questions   (A)

1. (RUSMO/2008/R4) The points M, N are on the sides AB, AC of $\triangle ABC$ respectively but not A, B, C, and $MC = AC, NB = AB$. P is the symmetric point of A with respect to the line BC. Prove that PA is angle bisector of the $\angle MPN$.

2. (CROATIA/2009) Given that CH is the altitude of the acute triangle ABC, O is the circumcircle of $\triangle ABC$ and T is the projection of C on the line AO. Prove that the line TH passes through the midpoint of the side BC.

3. (ITALY/2009) The acute $\triangle ABC$ is not equilateral, Γ is its circumcircle, the angle bisector of $\angle BAC$ intersects BC at K. The midpoint of the arc $\overset{\frown}{BC}$ (which contains A) is M. The line MK intersects Γ again at A'. The tangents to Γ at A and A' respectively intersect at T. The lines passing through A and A' and perpendicular to AK and AK' respectively intersect at R. Prove that T, R, K are collinear.

4. (BELARUS/2008) In the convex quadrilateral $ABCD$, $BC = CD$, $AB \neq AD$, $\angle BAC = \angle DAC$. The circle passing through A, C intersects AB, AD at M, N respectively. If $BN = a$, find DM.

5. (SMO/TST/2008) Let $\odot O$ be a circle, and let ABP be a line segment such that A, B lie on $\odot O$ and P is a point outside $\odot O$. Let C be a point on $\odot O$ such that PC is tangent to $\odot O$ and let D be the point on $\odot O$ such that CD is a diameter of $\odot O$ and intersects AB inside $\odot O$. Suppose that the lines DB and OP intersect at E. Prove that AC is perpendicular to CE.

6. (CMC/2008) In the given diagram, AB is the diameter of the semi-circle O, C is the midpoint of arc $\overset{\frown}{AB}$, M is the midpoint of the chord AC, $CH \perp BM$ at H. Prove that

    $$CH^2 = AH \cdot OH.$$

7. (CMC/2009) As shown in the right diagram, in the $\triangle ABC$, $AB > AC$, AE is the tangent line to the circumcircle of $\triangle ABC$ at A, D is on AB such that
    $$AD = AC = AE.$$
    Prove that the line segment DE passes through the incenter of $\triangle ABC$.

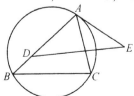

Testing Questions   (B)

1. (RUSMO/2008/R4) The inscribed circle ω of $\triangle ABC$ touches the sides BC, CA, AB at points A', B', C' respectively. The points K, L on ω satisfy

    $$\angle AKB' + \angle BKA' = \angle ALB' + \angle BLA' = 180°.$$

    Prove that the distances from the points A', B', C' to the line KL are equal.

2. (SNOVENIA/TST/2008-2009) $ABCD$ is a trapezium, CD, AB are the upper base and lower base respectively and $\angle ADC = 90°$, $AC \perp BD$. From D make $DE \perp BC$ at E. Prove that

    $$\frac{AE}{BE} = \frac{AC \cdot CD}{AC^2 - CD^2}.$$

3. (KOREA/2009) Let $\odot O$ be the circumcircle of an obtuse triangle ABC, where $\angle B$ is obtuse. the tangent line to $\odot O$ at C intersects the line AB at B_1. Let O_1 be the circumcenter of $\triangle AB_1C$. Take any point B_2 on the line segment BB_1 ($B_2 \neq B, B_1$), and let C_1 be the tangent point of the tangent from B_2 to $\odot O$ which is closer to C. Let O_2 be the circumcenter of $\triangle AB_2C_1$. Prove that the five points O, O_2, O_1, C_1, C are concyclic if $OO_2 \perp AO_1$.

4. (JAPAN/2009) Let $\varGamma$ be the circumcircle of $\triangle ABC$. The circle $\odot O$ touches the line segment BC at the point P and touches the arc $\overset{\frown}{BC}$ which does not have the point A at Q. If $\angle BAO = \angle CAO$, prove that $\angle PAO = \angle QAO$.

5. (IMO/Shortlist/2008) Given trapezoid $ABCD$ with parallel sides AB and CD, assume that there exist points E on line BC outside line segment BC, and F inside line segment AD such that $\angle DAE = \angle CBF$. Denote by I the point of intersection of CD and EF, and by J the point of intersection of AB and EF. Let K be the midpoint of line segment EF, assume it does not lie on line AB. Prove that I belongs to the circumcircle of ABK if and only if K belongs to the circumcircle of CDJ.

Lecture 13

Power of a Point with Respect to a Circle

The main findings on power of a point with respect to a circle are derived from the following theorems:

Theorem I. *Let $\odot O$ be a circle of radius R. For an interior point P of the circle, if AB is a chord passing through P, then the value of product $PA \cdot PB$ is independent of the choice of the chord passing through P. In particular, the constant value is given by taking $AB \perp OP$, so it is*

$$R^2 - OP^2.$$

Theorem II. *Let $\odot O$ be a circle of radius R. For a point P which is outside the circle or on the circumference, if PAB is a transversal line starting from P, intersecting the circle at A and B, then the value of $PA \cdot PB$ is independent of the choice of the transversal passing through P. In particular, the value is given by taking the tangent from P to the circle as the transversal, so it is*

$$OP^2 - R^2.$$

Consequence: When line segments AB and CD or their extensions intersect at P, then A, B, C, D are concyclic if and only if $AP \cdot PB = CP \cdot PD$.

It is clear the value $R^2 - OP^2$ (when P is inside the circle) or $OP^2 - R^2$ (when P is outside the circle) is determined by the distance of P from the circle, so it is defined as the **Power of the point** P **with respect to the circle** $\odot O$.

The *power of a point* has many applications in investigating questions about circles, where the concept of the radical axis is based on power of a point.

For any two non-concentric circles ω_1 and ω_2, the locus of all points with equal powers to the two circles is a line, called **radical axis** of the two circles.

Theorem III. *When the centers of three circles $\omega_1, \omega_2, \omega_3$ are not collinear, there exists exactly one point whose powers with respect to the three circles are equal. This point is said to be the **radical center** of the three circles.*

For given three circles, if any two have a radical axis, then the three radical axes may be parallel each other, or concurrent at their radical center.

Key Properties of radical axis:

- Two different concentric circles have no radical axis.

- The radical axis exists for any separated two non-concentric circles, and the radical axis is a line perpendicular to the line joining the centers of the two circles, passing through the midpoints of the four common tangent lines of the two circles.

- The radical axis exists for any two intersected circles, and it is the straight line passing through the two common points of the circles.

- The radical axis exists for any two circles if they are tangent each other, and it is the common tangent line at the common point of the circles.

Examples

Example 1. (BELARUS/2008) The pentagon $ABCDE$ is inscribed in a circle, the diagonals EC, AC intersect BD at the points L, K respectively. $BC = \sqrt{10}$. If the pints A, K, L, E are all on a circle Γ, find the length of the tangent from C to Γ.

Solution Let the length of the tangent from C to the circle Γ be a. Then

$$\angle BKA = \frac{1}{2}(\overset{\frown}{AB} + \overset{\frown}{CD}) \text{ and}$$

$$\angle BKA = \angle CEA = \frac{1}{2}(\overset{\frown}{AB} + \overset{\frown}{BC})$$

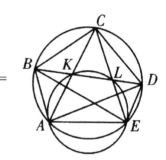

implies that $\overset{\frown}{BC} = \overset{\frown}{CD}$. Therefore $\angle CBK = \angle BAC$, so $\triangle BCK \sim \triangle ACB$ which gives

$$\frac{CK}{CB} = \frac{CB}{AC} \Rightarrow CK \cdot CA = CB^2.$$

The power of C to Γ gives $CK \cdot CA = a^2$, so

$$a = BC = \sqrt{10}.$$

Example 2. (HONG KONG/2006) O is the circumcenter of the convex quadrilateral $ABCD$. Given $AC \neq BD$, AC and BD intersect at E. If P is an interior point of the quadrilateral $ABCD$ such that $\angle PAB + \angle PCB = \angle PBC + \angle PDC = 90°$, prove that O, P, E are collinear.

Solution Let the circumcircles of the quadrilateral $ABCD$, $\triangle APC$ and $\triangle BPD$ be Γ, Γ_1 and Γ_2 respectively, then the radical axis of Γ and Γ_1 is the line AC, and the radical axis of Γ and Γ_2 is the line BD.

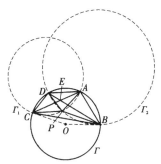

Since P is on both Γ_1 and Γ_2, so P is on the radical axis of Γ_1 and Γ_2.

Since BD and AC intersect at E, so E is the radical center of Γ, Γ_1 and Γ_2, therefore the line PE is the radical axis of Γ_1 and Γ_2.

It suffices to show that O has equal power to Γ_1 and Γ_2. For this note that

$$\angle APC = \angle PAB + \angle PCB + \angle ABC = 90° + \frac{1}{2}\angle AOC,$$

$$\angle ACO = \frac{1}{2}(180° - \angle AOC) = 180° - \left(90° + \frac{1}{2}\angle AOC\right)$$

$$= 180° - \angle APC = \angle ACP + \angle CAP,$$

therefore $\angle PCO = \angle CAP$, i.e., OC is tangent to Γ_1 at C. Similarly, OB is tangent to Γ_2 at B. Then $OC = OB$ implies that O has equal power to Γ_1 and Γ_2, so the conclusion is proven.

Example 3. (CMC/2009) In the given diagram, PA, PB are two tangents to the circle $\odot O$ at A and B. The line passing through P intersects $\odot O$ at C and D and intersects the chord AB at Q. Prove that

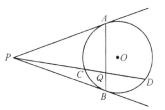

$$PQ^2 = PC \cdot PD - QC \cdot QD.$$

Solution Connect OA, OB, OP. Let the point of intersection of OP and AB be H, the midpoint of CD be M, connect OM. Then $OM \perp CD$ and $OH \perp AB$. Therefore Q, H, O, M are concyclic, hence

$$PQ \cdot PM = PH \cdot PO.$$

By applying the projection theorem to the Rt$\triangle APO$,

$$PA^2 = PH \cdot PO \Rightarrow PA^2 = PQ \cdot PM.$$

By considering the power of point P with respect to the circle $\odot O$, it follows that

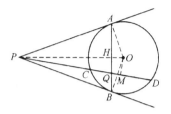

$$PQ \cdot PM = PA^2 = PC \cdot PD,$$

therefore

$$PQ(PQ + QM) = PC \cdot PD, \quad \text{so that} \quad PQ^2 = PC \cdot PD - PQ \cdot QM.$$

Since P, A, O, M and P, B, M, O are both concyclic, so P, B, M, O, A are concyclic. By considering the powers of Q with respect to this circle and $\odot O$, it follows that

$$PQ \cdot QM = QA \cdot QB = QC \cdot QD$$

since A, C, B, D are concyclic. Thus, $PQ^2 = PC \cdot PD - QC \cdot QD$.

Example 4. **(Euler's Theorem)** When the circumradius and inradius of $\triangle ABC$ are R and r respectively, and d is the distance between the circumcenter and the incenter, then

$$d^2 = R(R - 2r), \quad \text{or} \quad d = \sqrt{R(R - 2r)}.$$

Consequence: $R \geq 2r$, and $R = 2r$ if and only if $\triangle ABC$ is equilateral.

　　Solution As shown in the diagram below, let O and I be the circumcenter and incenter of $\triangle ABC$ respectively. Let the extension of AI intersect the $\odot O$ at D. Then D is the midpoint of the arc $\overparen{BC}$. Write

$$\alpha = \frac{1}{2}\angle A, \quad \beta = \frac{1}{2}\angle B,$$

then $\angle IBD = \angle IBC + \angle CBD = \beta + \alpha =$
$\angle IBA + \angle BAL = \angle BID$, therefore $\triangle DBI$ is isosceles with $DI = DB$.
By considering the power of I and the sine rule, it follows that

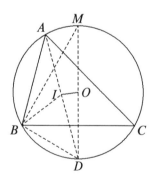

$$R^2 - d^2 = DI \cdot IA = 2R\sin\alpha \cdot \frac{r}{\sin\alpha} = 2Rr,$$

so that $d^2 = R^2 - 2Rr = R(R - 2r)$, as desired. The consequence is obvious.

Example 5. (CMO/2010) Two circles Γ_1 and Γ_2 intersect at A and B, as shown in the diagram below. A line passing through B intersects Γ_1 and Γ_2 at C and

D respectively, and another line passing through B intersects Γ_1 and Γ_2 at E and F respectively. The line CF intersects Γ_1 and Γ_2 at P and Q respectively. Let M and N be the midpoints of the arcs $\overset{\frown}{PB}$ and $\overset{\frown}{QB}$ respectively. Prove that C, F, M, N are concyclic if $CD = EF$.

Solution Connect $AC, AD, AE, AF, DF, CM, FN, AB$. Then

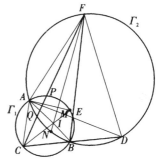

$\angle ADB = \angle AFB, \angle ACB = \angle AEF, CD = EF$
$\Rightarrow \triangle ACD \sim \triangle AEF \Rightarrow AD = AF$
$\Rightarrow \angle ADF = \angle AFD$
$\Rightarrow \angle ABC = \angle AFD = \angle ADF = \angle ABF$
$\Rightarrow AB$ is the angle bisector of $\angle CBF$.

Since $\overset{\frown}{PM} = \overset{\frown}{MB}$, so CM is the angle bisector of $\angle DCF$, and similarly, FN is the angle bisector of $\angle CFB$. Therefore BA, CM, FN are concurrent at the incenter I of $\triangle CFB$.

In view of the power of point I with respect to the circles Γ_1 and Γ_2, then

$$CI \cdot IM = AI \cdot IB = NI \cdot IF,$$

so C, F, M, N are concyclic.

Example 6. (APMO/2009) Let three circles Γ_1, Γ_2, Γ_3, which are non-overlapping and mutually external, be given in the plane. For each point P in the plane, outside the three circles, construct six points $A_1, B_1, A_2, B_2, A_3, B_3$ as follows: For each $i = 1, 2, 3$, A_i, B_i are distinct points on the circle Γ_i such that the lines PA_i and PB_i are both tangents to Γ_i. Call the point P *exceptional* if, from the construction, three lines $A_1 B_1, A_2 B_2, A_3 B_3$ are concurrent. Show that every exceptional point of the plane, if exists, lies on the same circle.

Solution Let O_i be the center and r_i the radius of circle Γ_i for each $i = 1, 2, 3$. Let P be an exceptional point, and let the three corresponding lines $A_1 B_1, A_2 B_2, A_3 B_3$ concur at Q. Construct the circle with diameter PQ. Call the circle Γ, its center O and its radius r. We now claim that all exceptional points lie on Γ.

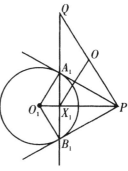

Let PO_1 intersect $A_1 B_1$ in X_1. As $PO_1 \perp A_1 B_1$, we see that X_1 lies on Γ. As PA_1 is a tangent to Γ_1, triangle $PA_1 O_1$ is right-angled and similar to triangle $A_1 X_1 O_1$. It follows that

$$\frac{O_1 X_1}{O_1 A_1} = \frac{O_1 A_1}{O_1 P}, \quad \text{i.e.,} \quad O_1 X_1 \cdot O_1 P = O_1 A_1^2 = r_1^2.$$

On the other hand, $O_1X_1 \cdot O_1P$ is also the power of O_1 with respect to Γ, so that

$$r_1^2 = O_1X_1 \cdot O_1P = (O_1O - r)(O_1O + r) = O_1O^2 - r^2, \qquad (*)$$

and hence

$$r^2 = OO_1^2 - r_1^2 = (OO_1 - r)(OO_1 + r).$$

Thus, r^2 is the power of O with respect to Γ_1. By the same token, r^2 is also the power of O with respect to Γ_2 and Γ_3. Hence O must be the radical center of the three given circles. Since r, as the square root of the power of O with respect to the three given circles, does not depend on P, it follows that all exceptional points lie on Γ.

Remark. In the event of the radical point being at infinity (and hence the three radical axes being parallel), there are no exceptional points in the plane, which is consistent with the statement of the problem.

Example 7. (USAMO/2009) Given circles ω_1 and ω_2 intersecting at points X and Y, let ℓ_1 be a line through the center of ω_1 intersecting ω_2 at points P and Q and let ℓ_2 be a line through the center of ω_2 intersecting ω_1 at points R and S. Prove that if P, Q, R and S lie on a circle then the center of this circle lies on line XY.

Solution

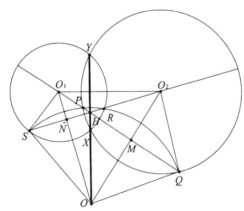

Let ω denote the circumcircle of P, Q, R, S and let O denote the center of ω. Line XY is the radical axis of circles ω_1 and ω_2. It suffices to show that O has equal power to the two circles; that is, to show that

$$OO_1^2 - O_1S^2 = OO_2^2 - O_2Q^2 \quad \text{or} \quad OO_1^2 + O_2Q^2 = OO_2^2 + O_1S^2.$$

Let M and N be the intersections of lines O_2O, ℓ_1 and O_1O, ℓ_2. Because circles ω and ω_2 intersect at points P and Q, we have $PQ \perp OO_2$ (or $\ell_1 \perp OO_2$), Hence

$$OO_1^2 - OQ^2 = (OM^2 + MO_1^2) - (OM^2 + MQ^2)$$
$$= (O_2M^2 + MO_1^2) - (O_2M^2 + MQ^2) = O_2O_1^2 - O_2Q^2$$

or

$$O_2O_1^2 + OQ^2 = OO_1^2 + O_2Q^2.$$

Likewise, we have $O_2O_1^2 + OS^2 = OO_2^2 + O_1S^2$. Because $OS = OQ$, we obtain that $OO_1^2 + O_2Q^2 = OO_2^2 + O_1S^2$, which is what to be proved.

Testing Questions   (A)

1. (CMC/2009) Given that the circle $\odot O$ touches line segment AB at M and touches the semi-circle of diameter AB at E. C is a point on the semi-circle such that $CD \perp AB$ at D and CD is the tangent line to $\odot O$ at point F. Connect CA, CM. Prove that

    (i) A, F, E are collinear;   (ii) $AC = AM$;   (iii) $MC^2 = 2MD \cdot MA$.

2. (THAILAND/2007) Given that the point P is outside the circle $\odot O$ and PA, PB are tangent to $\odot O$ at A and B respectively. M, N are the mid-points of the line segments AP and AB respectively. The extension of MN intersects $\odot O$ at C, where N is in between M and C. PC intersects $\odot O$ again at D, and the extension of ND intersects PB at Q. Prove that the quadrilateral $MNQP$ is a rhombus.

3. (ROMANIA/2008) Let ABC be a triangle and D, E, F are interior points of the line segments BC, CA, AB respectively, such that

    $$\frac{BD}{DC} = \frac{CE}{EA} = \frac{AF}{FB}.$$

    Prove that if the circumcenters of triangles DEF and ABC coincide, then ABC is equilateral.

4. (TURKEY/2008) A circle Γ and a line ℓ are given such that ℓ does not cut Γ. Determine the intersection set of the circles taking AB as diameter for all pairs of $\{A, B\}$ (lie on ℓ) and satisfy $P, Q, R, S \in \Gamma$ such that $PQ \cap RS = \{A\}$ and $PS \cap QR = \{B\}$.

5. (COLUMBIA/2009) In $\triangle ABC$, P is a point on the side BC, I_1, I_2 are in-centers of the $\triangle APB$ and $\triangle APC$ respectively. Γ_1, Γ_2 are circles passing through P and taking I_1, I_2 as the centers respectively. Let Q be the second point of intersection of Γ_1 and Γ_2. The points X_1, Y_1 are the points of intersection of Γ_1 with AB, BC respectively which are closer to B, and X_2, Y_2 are the points of intersection of Γ_2 with AC, BC respectively which are closer to C. Prove that the three lines X_1Y_1, X_2Y_2 and PQ are concurrent.

6. (BALTIC WAY/2007) The incircle of the triangle ABC touches the side AC at the point D. Another circle passes through D and touches the rays BC and BA, the latter at the point A. Determine the ratio $\dfrac{AD}{DC}$.

Testing Questions   (B)

1. (VIETNAM/2007) Let $ABCD$ be a trapezoid with the bigger base BC inscribed in the circle $\odot O$. Let P be a variable point on the line BC moving outside the line segment BC such that PA is not a tangent to the circle $\odot O$. A circle of diameter PD meets $\odot O$ at E ($E \neq D$). Denote by M the point of intersection of DE and BC, and by N the second point of intersection of PA and $\odot O$. Prove that the line MN passes through a fixed point.

2. (ITALY/TST/2008)) In the acute $\triangle ABC$, AM is the median on BC, BK, CL are altitudes, where the points M, K, L are on BC, CA, AB respectively. The line perpendicular to AM at A intersects the lines CL and BK at E and F respectively. Prove that

    (i) A is the midpoint of EF.

    (ii)   Let Γ be the circumcircle of $\triangle MEF$, and Γ_1, Γ_2 be two arbitrary circles which are tangent to line segment EF and the arc $\overset{\frown}{EF}$ that does not contain M. If Γ_1 and Γ_2 inteersect at P and Q, then M, P, Q are collinear.

3. (TURKEY/TST/2009) Quadrilateral $ABCD$ has an inscribed circle which centered at O with radius r. AB intersects CD at P; AD intersects BC at Q and the diagonals AC and BD intersects each other at E. If the distance from O to the line PQ is k, prove that $OE \cdot k = r^2$.

4. (BELARUS/2009) In the acute triangle ABC, $\angle C = 60°$, points B_1, A_1 are on the sides AC, BC respectively, and D is the second point of intersection of the circumcircles of $\triangle BCB_1$ and $\triangle ACA_1$. Prove that D is on the side AB if and only if $\dfrac{CB_1}{CB} + \dfrac{CA_1}{CA} = 1$.

Lecture 14

Some Important Theorems in Geometry

Theorem I. *(Menelaus' Theorem)*   *If a straight line cuts the sides AB, BC and CA (or their extensions) of a $\triangle ABC$ at points X, Y and Z respectively, then*

$$\frac{AX}{XB} \cdot \frac{BY}{YC} \cdot \frac{CZ}{ZA} = 1. \tag{14.1}$$

(Inverse Menelaus' Theorem)   For any given $\triangle ABC$, if X, Y, Z are points on lines AB, BC, CA respectively (where exact one point is on the extension of a side, or three points are all on the extensions of sides) such that (14.1) holds, then X, Y, Z must be collinear.

Theorem II. *(Ceva's Theorem)*   *For any given triangle ABC, let X, Y, Z be points with (i) all on the line segments BC, CA, AB; or (ii) exact one on one side and other two on the extensions of the two sides respectively. Then the lines AX, BY, CZ are parallel or concurrent if and only if*

$$\frac{BX}{XC} \cdot \frac{CY}{YA} \cdot \frac{AZ}{ZB} = 1. \tag{14.2}$$

Trigonometric Form of Ceva's Theorem   *The Condition (14.2) can be restated as*

$$\frac{\sin \angle BAX \cdot \sin \angle CBY \cdot \sin \angle ACZ}{\sin \angle CAX \cdot \sin \angle ABY \cdot \sin \angle BCZ} = 1. \tag{14.3}$$

Theorem III. *(Simson's Theorem)*   *For a $\triangle ABC$ and a point D which is outside the triangle, introduce three perpendicular lines from D to the sides BC, CA, intersecting them at A_1, B_1, C_1 respectively. Then A_1, B_1 and C_1 are collinear if and only if A, B, C, D are concyclic.*

Note: When A_1, B_1, C_1 are collinear, the line passing through them is called the **Simson line**.

101

Theorem IV. (Ptolemy's Theorem)   *Let $ABCD$ be a convex quadrilateral. Then*

$$AB \cdot CD + AD \cdot BC = AC \cdot BD \qquad (14.4)$$

if and only if A, B, C, D are concyclic.

   Extended Ptolemy's theorem   *For any convex quadrilateral $ABCD$, the inequality*

$$AB \cdot CD + AD \cdot BC \geq AC \cdot BD \qquad (14.5)$$

always holds, and the equality holds if and only if A, B, C, D are concyclic.

   Note: (14.5) is called the **Ptolemy's Inequality**.

Examples

Example 1. (Desargues' Theorem) For two triangles ABC and $A_1B_1C_1$ in a same plane, when the lines AA_1, BB_1, CC_1 are concurrent att a point S, then the three points P, Q, R must be collinear, where the lines BC and B_1C_1 intersect at P, the lines CA and C_1A_1 intersect at Q, and the lines AB and A_1B_1 intersect at R, as shown in the right diagram.

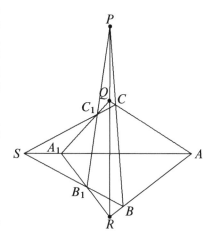

   Solution   Consider the line PC_1B_1 as a transversal to the $\triangle SBC$, by the Menelaus' theorem,

$$\frac{BP}{PC} \cdot \frac{CC_1}{C_1S} \cdot \frac{SB_1}{B_1B} = 1.$$

Since QC_1A_1 is a transversal to $\triangle SCA$, so

$$\frac{CQ}{QA} \cdot \frac{AA_1}{A_1S} \cdot \frac{SC_1}{C_1C} = 1.$$

   Similarly, considering RB_1A_1 as a transversal to $\triangle SAB$, then

$$\frac{AR}{RB} \cdot \frac{BB_1}{B_1S} \cdot \frac{SA_1}{A_1A} = 1.$$

Multiplying the three equalities gives

$$\frac{BP}{PC} \cdot \frac{CQ}{QA} \cdot \frac{AR}{RB} = 1.$$

Since P, Q, R are all at the extensions of BC, CA, AB respectively, P, Q, R must be collinear by the inverse Menelaus' theorem.

Example 2. (BULGARIA/2009) The inscribed circle $\odot I$ of $\triangle ABC$ touches the sides BC, CA, AB at A_1, B_1, C_1 respectively. Let l be a line passing through I and A', B', C' be the symmetric points of A_1, B_1, C_1 with respect to l respectively. Prove that the lines AA', BB', CC' are concurrent.

Solution Use $d_a(X), d_b(X), d_c(X)$ to denote the distances of point X from the lines BC, CA, AB respectively. The Ceva's theorem in trigonometric form gives

AA', BB', CC' are collinear $\Leftrightarrow$

$$\frac{\sin \angle CAA'}{\sin \angle BAA'} \cdot \frac{\sin \angle ABB'}{\sin \angle CBB'} \cdot \frac{\sin \angle BCC'}{\sin \angle ACC'} = 1$$

$$\Leftrightarrow \frac{d_b(A')/AA'}{d_c(A')/AA'} \cdot \frac{d_c(B')/BB'}{d_a(B')/BB'} \cdot \frac{d_a(C')/CC'}{d_b(C')/CC'} = 1$$

$$\Leftrightarrow \frac{d_b(A')}{d_c(A')} \cdot \frac{d_c(B')}{d_a(B')} \cdot \frac{d_a(C')}{d_b(C')} = 1.$$

Since $A'B_1 = A_1B'$ and CB, CA are both tangent to $\odot I$, so, in degrees,

$$\angle B'A_1C = \tfrac{1}{2} \overparen{B'A_1} = \tfrac{1}{2} \overparen{A'B_1} = \angle A'B_1A,$$
$$\Rightarrow d_a(B') = A_1B' \sin \angle B'A_1C = B_1A' \sin \angle A'B_1A = d_b(A').$$

Similarly, $d_b(C') = d_c(B')$ and $d_c(A') = d_a(C')$, the conclusion thus is proven.

Example 3. (**Carnot's Theorem**) Let P be a point on the circumcircle of $\triangle ABC$. From P introduce three lines to intersect the lines BC, CA, AB at D, E, F respectively, such that $\angle PDB = \angle PEC = \angle PFB$. Then D, E, F are collinear.

Solution It suffices to show that $\angle PDF + \angle PDE = 180°$. Since $\angle PDB = \angle PFB$ implies that $BPDF$ is cyclic, so

$$\angle PDF + \angle PBF = 180°.$$

$ABPC$ is cyclic implies that

$$\angle PBF = \angle PCE,$$

and $\angle PDE = \angle PCE$ implies $PDCE$ is cyclic, so

$$\angle PDE = \angle PCE = \angle PBF,$$

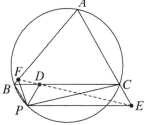

thus, $\angle PDF + \angle PDE = \angle PDF + \angle PBF = 180°$, so D, E, F are collinear.

Example 4. (USAMO/2010) Let $AXYZB$ be a convex pentagon inscribed in a semicircle of diameter AB. Denote by P, Q, R, S the feet of the perpendiculars from Y onto lines AX, BX, AZ, BZ, respectively. Prove that the acute angle formed by lines PQ and RS is half the size of $\angle XOZ$, where O is the midpoint of line segment AB.

Solution   As shown in the diagram below,
Let T be the foot of the perpendicular from Y to AB, then, by the Simson's theorem, P, Q, T are collinear and so are S, R, T.
Since

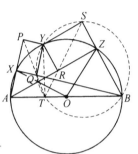

$$\angle YQB = \angle YTB = \angle YSB = 90°,$$

S, Y, Q, T, B are concyclic, so that

$$\angle PTS = \angle QTS = \angle QBS = \angle XBZ = \frac{1}{2}\angle XOZ.$$

Example 5. (CMC/2008) Given a convex quadrilateral $ABCD$ with $\angle B + \angle D < 180°$, P is a variable point in the same plane. Let $f(P) = PA \cdot BC + PD \cdot CA + PC \cdot AB$. Prove that P, A, B, C are concyclic if $f(P)$ reaches its minimum value.

Solution   As shown in the right diagram, for any point P in the plane, the Ptolemy's inequality gives

$$PA \cdot BC + PC \cdot AB \geq PB \cdot AC,$$

therefore

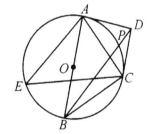

$$\begin{aligned} f(P) &= PA \cdot BC + PC \cdot AB + PD \cdot CA \\ &\geq PB \cdot CA + PD \cdot CA \\ &= (PB + PD) \cdot CA \geq BD \cdot CA, \end{aligned}$$

where the equality holds if and only if $PABC$ is cyclic. Therefore $f(P)$ has a lower bound $BD \cdot AC$.
However, this bound is reachable when and only when P is the point of intersection of the circumcircle of $\triangle ABC$ and line segment BD, so P, A, B, C are concyclic.

Example 6. (Pascal's Theorem) If a hexagon $ABCDEF$ is inscribed in a circle such that the lines AB and ED produced intersect at H, the lines BC and FE produced intersect at K, and the lines AF and CD produced intersect at I, prove that H, K, I are collinear.

Solution Suppose that the lines $AB, CD,$ EF form a $\triangle XYZ$. Considering the lines AFI, BCK, HDE as three transversals of $\triangle XYZ$, then the Menelaus' Theorem yields

$$\frac{XA}{AZ} \cdot \frac{ZF}{FY} \cdot \frac{YI}{IX} = 1,$$

$$\frac{XB}{BZ} \cdot \frac{ZK}{KY} \cdot \frac{YC}{CX} = 1,$$

$$\frac{XH}{HZ} \cdot \frac{ZE}{EY} \cdot \frac{YD}{DX} = 1.$$

Multiplying the three equalities up and considering

$$XA \cdot XB = XC \cdot XD, \quad YC \cdot YD = YE \cdot YF$$

and $ZE \cdot ZF = ZA \cdot ZB$, we obtain $\dfrac{YI}{IX} \cdot \dfrac{XH}{HZ} \cdot \dfrac{ZK}{KY} = 1$. By the inverse Menelaus' Theorem, I, K, H are collinear.

Example 7. (Pappus' Theorem) Given that A, B, C are three points on a line ℓ_1 and a, b, c are three points on another line ℓ_2. If the line segments Ab, Ba intersect at N, the line segments Ac, Ca intersect at L and the line segments Bc, Cb intersect at M respectively, prove that N, L, M are collinear.

Solution Consider the triangle UVW formed by the lines Ab, aC, Bc, as shown in the given right diagrm.
There are five transversals:

$$ALc, \quad CMb, \quad aNB, \quad CBA, \quad abc.$$

Applying the Menelaus' theorem then yields

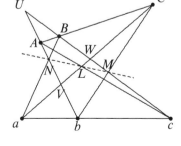

$$\frac{VL}{LW} \cdot \frac{Wc}{cU} \cdot \frac{UA}{AV} = 1,$$

$$\frac{VC}{CW} \cdot \frac{WM}{MU} \cdot \frac{Ub}{bV} = 1,$$

$$\frac{Va}{aW} \cdot \frac{WB}{BU} \cdot \frac{UN}{NV} = 1,$$

$$\frac{VC}{CW} \cdot \frac{WB}{BU} \cdot \frac{UA}{AV} = 1 \quad \text{and} \quad \frac{Va}{aW} \cdot \frac{Wc}{cU} \cdot \frac{Ub}{bV} = 1.$$

By multiplying the first three equalities and then divided the resultant equality by the product of the last two equalities, it is obtained that

$$\frac{VL}{LW} \cdot \frac{WM}{MU} \cdot \frac{UN}{NV} = 1.$$

Thus, by the inverse Menelaus' theorem, L, N, M are collinear.

Example 8. (**Newton Line**) $ABCD$ is a quadrilateral such that the rays BA and CD intersect at E, and the rays AD and BC intersect at F. Let N, L, M be the midpoints of EF, AC, BD respectively, prove that N, L, M are collinear.

Note: The line passing through N, L, M is called the *Newton line* of the quadrilateral $ABCD$.

Solution Let the midpoints of EB, EC, BC be P, Q, R respectively. Then the midpoint theorem indicates that
Q, L, R are collinear, and

$$\frac{QL}{LR} = \frac{EA}{AB}.$$

Similarly, P, M, R are collinear, and

$$\frac{RM}{MP} = \frac{CD}{DE}.$$

Besides, N, Q, P are collinear and

$$\frac{PN}{NQ} = \frac{BF}{FC}.$$

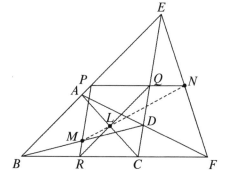

Applying the Menelaus' theorem to $\triangle EBC$ and the transversal ADF then gives

$$\frac{EA}{AB} \cdot \frac{BF}{FC} \cdot \frac{CD}{DE} = 1,$$

therefore $\dfrac{QL}{LR} \cdot \dfrac{RM}{MP} \cdot \dfrac{PN}{NQ} = \dfrac{EA}{AB} \cdot \dfrac{CD}{DE} \cdot \dfrac{BF}{FC} = 1$. By applying the inverse Menelaus' theorem to the $\triangle PQR$ and the three points M, L, N, the conclusion that M, L, Q are collinear is proven.

Testing Questions   (A)

1. (CMC/2009) Given that line segments AB, CD, EF are three non-intersected chords of a circle. When three quadrilaterals are formed by taking any two chords as a pair of opposite sides, and let M, N, P be the intersection points of two diagonals for these quadrilaterals, prove that M, N, P are collinear.

2. (HUNGARY/2007-2008) It is known that all the angles of the convex hexagon $A_1 A_2 A_3 A_4 A_5 A_6$ are obtuse, the circle Γ_i, $(1 \leq i \leq 6)$ has the center A_i and such that Γ_i is tangent to the circles Γ_{i-1} and Γ_{i+1} externally, where $\Gamma_0 = \Gamma_6, \Gamma_1 = \Gamma_7$.

    Let the line passing through the two tangent points on Γ_1 and the line passing through the two tangent points on Γ_3 intersect at a point, and the line joining the point and A_2 is e. Similarly, let the line defined by the circles Γ_3, Γ_5 and A_4 be f, and the line defined by the circles Γ_5, Γ_1 and A_6 be g. Prove that the three lines e, f, g are concurrent.

3. (INDIA/TST/2008) Let $\triangle ABC$ is not isosceles, Γ is its inscribed circle, touching the three sides BC, CA, AB at D, E, F respectively. If the lines FD, DE, EF intersect the lines CA, AB, BC at the points U, V, W respectively, and the midpoints of DW, EU, FV are L, M, N respectively, prove that L, M, N are collinear.

4. (CMC/2008) Given the diameter of circumcircle of $\triangle ABC$ is 25, the lengths of AB, BC, CA are all positive integers with $AB > BC$, and the distances from the circumcenter O to the sides AB, BC are both positive integers, find the lengths of AB, BC, CA.

5. (ESTONIA/TST/2009) Given that A', B', C' are on the sides BC, CA, AB of $\triangle ABC$ respectively, satisfying $\dfrac{BA'}{A'C} = \dfrac{CB'}{B'A} = \dfrac{AC'}{C'B}$. By passing through A' introduce a line l such that $l \parallel B'C'$. If l intersects the lines AC, AB at P, Q respectively, prove that $\dfrac{PQ}{B'C'} \geq 2$.

6. (CMC/2009) A pair of opposite interior angles, say $\angle A$ and $\angle C$, of the cyclic quadrilateral $ABCD$ is partitioned by the diagonal AC as $\angle A = \alpha_1 + \alpha_2, \angle C = \alpha_3 + \alpha_4$. Prove that

    $$\sin(\alpha_1 + \alpha_2) \sin(\alpha_2 + \alpha_3) \sin(\alpha_3 + \alpha_4) \sin(\alpha_4 + \alpha_1)$$
    $$\geq 4 \sin \alpha_1 \sin \alpha_2 \sin \alpha_3 \sin \alpha_4.$$

Testing Questions   (B)

1. (CMC/2010) Let O be the circumcenter of the acute triangle ABC, K a point on BC (but it is not the midpoint of BC), D a point on the extension of AK. The lines BD and AC intersect at N and the lines AB and CD intersect at M. Prove that if $OK \perp MN$ then A, B, D, C are concyclic.

2. (**Steiner-Miquel Theorem**) $ABCD$ is a convex quadrilateral, and the extensions of AB, DC intersect at E, the extensions of AD, BC intersect at F. Prove that the four circumcircles of triangles BCE, CDF, ADE, ABF are concurrent at one point.

3. (BALKAN MO/2009) Let MN be a line parallel to the side BC of a triangle ABC, with M on side AB and N on side AC. The lines BN and CM meet at point P. The circumcircles of triangles BMP and CNP meet at two distinct points P and Q. Prove that $\angle BAQ = \angle CAP$.

4. (CHNMO/TST/2010) $ABCD$ is a convex quadrilateral, and the extensions of AB, DC intersect at E, the extensions of AD, BC intersect at F. The circumcircles of $\triangle BEC$ and $\triangle CFD$ intersect at C and P. Prove that $\angle BAP = \angle CAD$ if and only if $BD \parallel EF$.

Lecture 15

Five Centers of a Triangle

Theorem I. *The three medians of a triangle intersect at one common point, denoted by G, and each median is partitioned by G as two parts of ratio $2 : 1$. The common point G is called* **center of gravity** *or* **centroid** *of the triangle.*

Consequence *An interior point P of $\triangle ABC$ is the center of gravity of $\triangle ABC$ if and only if*

$$[PBC] = [PCA] = [PAB].$$

(Here the notation of $[XYZ]$ denotes the area of $\triangle XYZ$.)

Theorem II. *For any triangle, the perpendicular bisectors of three sides intersect at a common point O. The O is the center of circumcircle of the triangle, called* **circumcenter** *of the triangle.*

Theorem III. *The three altitudes of any $\triangle ABC$ intersect at one common point H, called* **orthocenter** *of the triangle.*

Theorem IV. *For any triangle, its angle bisectors of three interior angles intersect at one common point, denoted by I as usual, called* **incenter** *(or* **inner center***) of the triangle. I is the center of inscribed circle of the triangle.*

Theorem V. *For a triangle, the angle bisectors of one interior angle and two exterior angles of the other two interior angles intersect at common point, called* **excenter** *of the triangle. There are three such points for a triangle, and each is the center of an escribed circle of the triangle.*

Note: For an isosceles triangle, its center of gravity, circumcenter, orthocenter and incenter are all on the symmetric axis of the triangle. For an equilateral triangle, the above four centers coincide to one point, called **center** of the equilateral triangle.

Examples

Example 1. For a given $\triangle ABC$ with G as its center of gravity and any point M in the same plane, the following equality always hold:

$$MA^2 + MB^2 + MC^2 = GA^2 + GB^2 + GC^2 + 3MG^2.$$

Note: The conclusion gives another definition of the center of gravity of the triangle ABC: A point G is the center of gravity of $\triangle ABC$ if and only if

$$GA^2 + GB^2 + GC^2 = \min\{MA^2 + MB^2 + MC^2\},$$

where M can be any point in the same plane.

   Solution   As shown in the figure, let G be the center of gravity and the three medians be AD, BE, CF respectively. Use K to denote the mid-point of AG and connect MK, MD. Then $AK = KG = GD$ and

$$MB^2 + MC^2 = 2(BD^2 + MD^2), \quad (15.1)$$
$$MA^2 + MG^2 = 2(MK^2 + GK^2), \quad (15.2)$$
$$2(MD^2 + MK^2) = 4(MG^2 + GK^2). \quad (15.3)$$

By $(15.1) + (15.2) + (15.3)$,

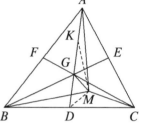

$$MA^2 + MB^2 + MC^2 = 3MG^2 + 2(BD^2 + GK^2) + 4GK^2.$$
$$\because \ 4GK^2 = GA^2,$$
$$2(BD^2 + GK^2) = 2(BD^2 + GD^2) = GB^2 + GC^2,$$
$$\therefore \ MA^2 + MB^2 + MC^2 = GA^2 + GB^2 + GC^2 + 3GM^2.$$

Example 2. (Euler line) For any triangle ABC, its center of gravity G, circumcenter O and orthocenter H are collinear, and G is between O and H such that $HG = 2OG$.

   Solution   If A' and C' are the midpoints of BC and AB respectively, then G is on AA' and $AG = 2GA'$. Connect OA', OG. Let H' be on the extension of OG such that $H'G = 2GO$, then $\triangle AGH' \sim \triangle A'OG$, i.e., $AH' \parallel OA'$, hence $AH' \perp BC$ since $OA' \perp BC$.

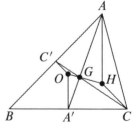

Similarly, $CH' \perp AB$.
Thus, H' is just the orthocenter H. The conclusion is proven.

Example 3. H is the orthocenter of an acute $\triangle ABC$ and $AH = p, BH = q, CH = r$. Show that $aqr + brp + cpq = abc$.

Solution Below we show the conclusion by using trigonometric method.

Let $\dfrac{a}{\sin A} = \dfrac{b}{\sin B} = \dfrac{c}{\sin C} = 2R$, where R is the circumradius of $\triangle ABC$. Then

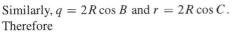

$$\angle AHE = 90° - \angle DAC = \angle C$$
$$\Rightarrow p = \frac{AE}{\sin C} = \frac{AB \cos A}{\sin C} = 2R \cos A.$$

Similarly, $q = 2R \cos B$ and $r = 2R \cos C$.
Therefore
$$aqr + brp + cpq = abc$$
$$\Leftrightarrow 2R \sin A \cdot 2R \cos B \cdot 2R \cos C + 2R \sin B \cdot 2R \cos C \cdot 2R \cos A$$
$$+ 2R \sin C \cdot 2R \cos A \cdot 2R \cos B = 2R \sin A \cdot 2R \sin B \cdot 2R \sin C$$
$$\Leftrightarrow \sin A \cos B \cos C + \sin B \cos C \cos A + \sin C \cos A \cos B = \sin A \sin B \sin C$$
$$\Leftrightarrow \tan A + \tan B + \tan C = \tan A \tan B \tan C,$$

and the last equality is well known as a basic property of a triangle.

Example 4. (CMC/2008) In an acute triangle ABC, D, E, F are the midpoints of the sides BC, CA, AB respectively. On the extensions of EF, FD, DE take points P, Q, R respectively, such that $AP = BQ = CR$. Prove that the circumcenter of $\triangle PQR$ is orthocenter of the $\triangle ABC$.

Solution Let AL, BM, CN be the altitudes of $\triangle ABC$, and H be the orthocenter. Suppose that EF intersects AL at K. Then

$$AP^2$$
$$= PK^2 + AK^2 = PH^2 - KH^2 + AK^2$$
$$= PH^2 (AK + KH)(AK - KH)$$
$$= PH^2 + AH \cdot HL.$$

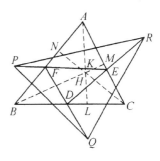

Similarly,

$$BQ^2 = QH^2 + BH \cdot HM \quad \text{and}$$
$$CR^2 = RH^2 + CH \cdot HN.$$

Since $AP = BQ = CR$ and $ABLM, BCMN$ are cyclic, therefore

$$AH \cdot HL = BH \cdot HM = CH \cdot HN,$$

so that $PH = QH = RH$, hence H is the circumcenter of $\triangle PQR$.

Example 5. (CMC/2009) In $\triangle ABC$, the points D, E are on AB, AC respectively such that $DE \parallel BC$. The inscribed circle of $\triangle ADE$ touches DE at M, the escribed circle on side BC of $\triangle ABC$ touches BC at N. BE and CD intersect at P. Prove that M, N, P are collinear.

Solution Suppose that BE and MN intersect at P'. Since $DE \parallel BC$,

$$\frac{BP}{PE} = \frac{BC}{DE} \text{ and } \frac{BP'}{P'E} = \frac{BN}{EM}, \text{ it suffices to show that}$$

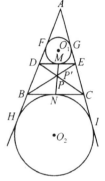

$$\frac{BN}{EM} = \frac{BC}{DE} \text{ or } \frac{BN}{BC} = \frac{EM}{DE}.$$

Let O_1, O_2 be centers of the inscribed circle and escribed circle respectively, and F, G, H, I be the related tangent points, as shown in the right diagram. Then

$$\begin{aligned}
EM &= \tfrac{1}{2}(AE + DE - AD), \\
AH &= AB + BH = AB + BN \\
&= AI = \tfrac{1}{2}(AB + BC + AC),
\end{aligned}$$

so that $BN = AH - AB = \tfrac{1}{2}(AC + BC - AB)$. Since $\triangle ADE \sim \triangle ABC$, so it is possible to let

$$\frac{AB}{AD} = \frac{BC}{DE} = \frac{AC}{AE} = k.$$

Then

$$\begin{aligned}
\frac{BN}{BC} &= \frac{\tfrac{1}{2}(AC + BC - AB)}{BC} = \frac{k(AE + DE - AD)}{2k \cdot DE} \\
&= \frac{AE + DE - AD}{2DE} = \frac{EM}{DE}.
\end{aligned}$$

Thus, the conclusion is proven.

Example 6. (CMC/2010) The inscribed circle of $\triangle ABC$ has center I and touches sides AC, AB at points E, F respectively. Let M be a point on line segment EF. Prove that the areas of $\triangle MAB$ and $\triangle MAC$ are equal if and only if $MI \perp BC$.

Solution Introduce $MP \perp AC$ at P and $MQ \perp AB$ at Q. Suppose that $\odot I$ touches BC at D, then $ID \perp BC$, $IF \perp AB$ and $IE \perp AC$.

$AF = AE$ implies that $\angle AFM = \angle AEM$, hence $\triangle QFM \sim \triangle PEM$, so that $\dfrac{MQ}{MP} = \dfrac{MF}{ME}$. Since

$$\frac{[MAB]}{[MAC]} = \frac{MQ \cdot AB}{MP \cdot AC} = \frac{MF}{ME} \cdot \frac{AB}{AC},$$

so $[MAB] = [MAC]$ if and only if

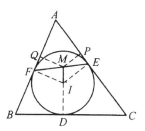

$$\frac{AB}{AC} = \frac{ME}{MF}. \qquad (15.4)$$

Below is to prove that (15.4) holds if and only if $MI \perp BC$.

When $MI \perp BC$, then M is on the line ID. $BDIF$ and $CDIE$ both are cyclic implies that

$$\angle MIF = \angle B, \qquad \angle MIE = \angle C.$$

Then the applications of sine rule to $\triangle MIF$ and $\triangle MIE$ give

$$\frac{MF}{\sin \angle MIF} = \frac{FI}{\sin \angle IMF} = \frac{IE}{\sin \angle EMI} = \frac{ME}{\sin \angle MIE},$$

namely $\dfrac{ME}{MF} = \dfrac{\sin \angle C}{\sin \angle B} = \dfrac{AB}{AC}$ due to sine rule again.

Conversely, when $\dfrac{AB}{AC} = \dfrac{ME}{MF}$, suppose that the line ID intersects EF at M', then above proof indicates that $\dfrac{M'E}{M'F} = \dfrac{AB}{AC}$, so $\dfrac{ME}{MF} = \dfrac{M'E}{M'F}$ which implies that $M' = M$. Thus, M is on the line ID, i.e., $IM \perp BC$.

Example 7. (APMO/2007) Let ABC be an acute angled triangle with $\angle BAC = 60°$ and $AB > AC$. Let I be the incenter, and H the orthocenter of the triangle ABC. Prove that

$$2\angle AHI = 3\angle ABC.$$

Solution Let D be the intersection point of the lines AH and BC. Let K be the intersection point of the circumcircle O of the triangle ABC and the line AH. Let the line through I perpendicular to BC meet BC and the minor arc BC of the circumcircle O at E and N, respectively. We have

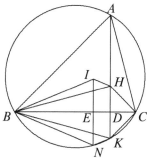

$$
\begin{aligned}
\angle BIC &= 180° - (\angle IBC + \angle ICB) \\
&= 180° - \tfrac{1}{2}(\angle ABC + \angle ACB) \\
&= 90° + \tfrac{1}{2}\angle BAC = 120°, \\
\angle BNC &= 180° - \angle BAC = 120° = \angle BIC.
\end{aligned}
$$

Since $IN \perp BC$, we have $IE = EN$ (otherwise, say $IE < EN$, then $\angle IBC < \angle NBC$ and $\angle ICB < \angle NCB$ but $\angle IBC + \angle ICB = \angle NBC + \angle NCB = 60°$, a contradiction).

Now, since H is the orthocenter of the triangle ABC, $HD = DK$. Also because $ED \perp IN$ and $ED \perp HK$, we conclude that $IHKN$ is an isosceles trapezoid with $IH = NK$.

Hence

$$\angle AHI = 180° - \angle IHK = 180° - \angle AKN = \angle ABN.$$

Since $IE = EN$ and $BE \perp IN$, the triangles IBE and NBE are congruent. Therefore

$$\angle NBE = \angle IBE = \angle IBC = \angle IBA = \frac{1}{2}\angle ABC$$

and thus

$$\angle AHI = \angle ABN = \frac{3}{2}\angle ABC.$$

Example 8. (TURKEY/2008) Given an acute triangle ABC, O is the circumcenter and H is the orthocenter. Let A_1, B_1, C_1 be the midpoints of the sides BC, CA and AB respectively. Rays HA_1, HB_1, HC_1 cut the circumcircle of $\triangle ABC$ at A_0, B_0 and C_0 respectively. Prove that O, H and H_0 are collinear if H_0 is the orthocenter of $\triangle A_0 B_0 C_0$.

    Solution   Connect HB, HC, A_0B, A_0C, A_0A. The orthocenter of $\triangle ABC$ is H implies that

$$\begin{aligned}\angle BHC &= 180° - (90° - \angle B) - (90° - \angle C)\\ &= \angle B + \angle C = 180° - \angle A = \angle BA_0C.\end{aligned}$$

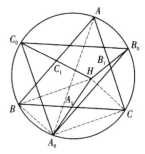

Since HA_0 passes through the midpoint A_1 of BC, so the quadrilateral $BHCA_0$ is a parallelogram, hence

$$\angle ACA_0 = 90°.$$

Therefore AA_0 is a diameter of the circumcircle circle $\odot O$. Similarly, CC_0, BB_0 are also diameters of $\odot O$.

    Thus, The triangles ABC and $A_0B_0C_0$ are symmetric with respect to O, hence H and H_0 are symmetric in O, i.e., O is the midpoint of line segment HH_0. The conclusion is proven.

Testing Questions   (A)

1. Let O, H be the circumcenter and orthocenter of an acute triangle ABC respectively. Prove that the maximal value of $[AOH], [BOH], [COH]$ (the notation $[XYZ]$ denotes area of $\triangle XYZ$) is equal to sum of the other two.

2. (CMC/2008) The circle taking side BC of $\triangle ABC$ as the diameter intersects the lines AB, AC at points E, F respectively. The tangent lines at E, F intersect at P. The line AP intersects the line BF at a point D. Prove that D, C, E are collinear.

3. (AUSTRIA/2009) Let D, E, F be the midpoints of the sides BC, CA, AB of $\triangle ABC$, H_a, H_b, H_c be the perpendicular feet of the three sides of $\triangle ABC$ respectively, and P, Q, R be the the midpoints of sides $H_b H_c, H_c H_a, H_a H_b$ of $\triangle H_a H_b H_c$. Prove that the lines PD, QE, RF are concurrent.

4. (IRAN/TST/2009) In triangle ABC, D, E and F are the points of tangency of incircle with the center of I to BC, CA and AB respectively. Let M be the feet of perpendicular from D to EF and P is on DM such that $DP = MP$. If H is the orthocenter of BIC, prove that PH bisects EF.

5. (BULGARIA/TST/2009) The three escribed circles of $\triangle ABC$ touch the line segments AB, BC, CA at the points M, N, P respectively. I and O are the incenter and circumcenter of $\triangle ABC$ respectively. Prove that if the quadrilateral $AMNP$ is cyclic, then

    (i) M, P, I are collinear;

    (ii) I, O, N are collinear.

6. (CMC/2008) It is known that the circumradius R of the acute triangle ABC is 1, $\angle BAC = 60°$, the orthocenter and circumcenter of $\triangle ABC$ are H and O respectively. Line OH intersects the extension of BC at the point P.

    (i)   Find area of the concave quadrilateral $ABHC$;

    (ii)   Find the value of $PO \cdot OH$.

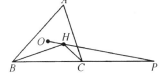

7. (GREECE/TST/2009) The centroid and circumcenter of $\triangle ABC$ are denoted by G and O respectively. If the perpendicular bisectors of GA, GB, GC intersect pairwise at the points A_1, B_1, C_1 respectively, prove that O is the centroid of $\triangle A_1 B_1 C_1$.

Testing Questions   (B)

1.   (IMO/2009) Let ABC be a triangle with circumcenter O. The points P and Q are interior points of the sides CA and AB, respectively. The circles k passes through the midpoints of the line segments BP, CQ and PQ. Prove that if the line PQ is tangent to circle k then $OP = OQ$.

2.   (CWMO/2009) D is an interior point of the side BC of an acute triangle ABC, the circle taking BD as its diameter intersects the lines AB, AD at X, P respectively, where P is neither B nor D. The circle taking CD as its diameter intersect the lines AC, AD at Y (different from C) and Q (different from D) respectively. From A introduce $AM \perp PX$ at M and $AN \perp QY$ at N.

Prove that $\triangle AMN \sim \triangle ABC$ if and only if the line AD passes through the circumcenter of $\triangle ABC$.

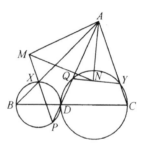

3.   (RUSMO/2010) The lines tangent to circle ω at points A and B intersect at point O. The point I is the center of ω. On the minor arc $\overset{\frown}{AB}$, a point C is chosen not on the midpoint of the arc. Lines AC and OB intersect at point D. Lines BC and OA intersect at point E. Prove that the circumcenters of triangles ACE, BCD, and OCI are collinear.

4.   (BELARUS/2009) Given that the diagonals AC and BD of a convex quadrilateral $ABCD$ intersect at T. the orthocenter of $\triangle ABT$ cincides with the circumcenter of $\triangle CDT$. Prove that

   (i)   the convex quadrilateral $ABCD$ is cyclic;

   (ii)   the circumcenter of $\triangle CDT$ is on the circumcircle of the quadrilateral $ABCD$.

5.   (CMO/TST/2008) Given that ABC is a triangle, and a line l cuts the lines BC, CA, AB at D, E, F respectively. O_1, O_2, O_3 are the circumcenters of $\triangle AEF, \triangle BFD, \triangle CDE$ respectively. Prove that the orthocenter H of $\triangle O_1O_2O_3$ is on the line l.

Solutions to Testing Questions

Solutions to Testing Questions

Solutions to Testing Questions   1

Testing Questions   (1-A)

1. The application of long divisions changes the given equation to

$$x + 2 + \frac{x - 2}{x^2 + 2x - 3} = x + 2 + \frac{x - 2}{2x^2 + x + 1},$$

so

$$(x-2)[(2x^2+x+1)-(x^2+2x-3)] = 0, \quad \text{i.e. } (x - 2)(x^2 - x + 4) = 0.$$

Since $x^2 - x + 4 > 0$ for all real x, $x = 2$ is the unique real root.

2. From the given equation,

$$1 - \frac{1}{x + 2} + 1 - \frac{1}{x + 7} = 1 - \frac{1}{x + 3} + 1 - \frac{1}{x + 6},$$

$$\frac{1}{x + 6} - \frac{1}{x + 7} = \frac{1}{x + 2} - \frac{1}{x + 3},$$

$$(x + 2)(x + 3) = (x + 6)(x + 7),$$

$$x^2 + 5x + 6 = x^2 + 13x + 42,$$

$$x = -\frac{9}{2}.$$

3. The value of x must be such that all the denominators in given equation are not zero,

$$\frac{1}{(x + k)(x + k + 1)} = \frac{1}{x + k} - \frac{1}{x + k + 1} \text{ for } k = -1, 0, 1, \cdots, 9,$$

119

the equation becomes

$$\frac{1}{x-1} - \frac{1}{x+10} = \frac{11}{12},$$
$$12[(x+10)-(x-1)] = 11(x-1)(x+10),$$
$$x^2 + 9x - 22 = 0,$$
$$(x+11)(x-2) = 0,$$

therefore $x_1 = 2$, $x_2 = -11$.

4. The substitution $y = x^2 - 8$ gives

$$\frac{1}{y+11x} + \frac{1}{y+2x} + \frac{1}{y-13x} = 0,$$
$$(y+2x)(y-13x) + (y+11x)(y-13x) + (y+11x)(y+2x) = 0,$$
$$3y^2 - 147x^2 = 0,$$
$$y = \pm 7x.$$

When $y = -7x$, then

$$x^2 + 7x - 8 = 0 \Rightarrow (x+8)(x-1) = 0, \quad x_1 = -8, \quad x_2 = 1.$$

When $y = 7x$, then

$$x^2 - 7x - 8 = 0 \Rightarrow (x-8)(x+1) = 0, \quad x_3 = 8, \quad x_4 = -1.$$

Thus, $x = \pm 1$ or ± 8.

5. By completing square,

$$\left(x - \frac{x}{x+1}\right)^2 + 2 \cdot \frac{x^2}{x+1} = \frac{5}{4},$$
$$4\left(\frac{x^2}{x+1}\right)^2 + 8 \cdot \frac{x^2}{x+1} - 5 = 0.$$

Let $y = \dfrac{x^2}{x+1}$, then $4y^2 + 8y - 5 = 0$, so $(2y-1)(2y+5) = 0$, i.e.
$y = \dfrac{1}{2}$ or $-\dfrac{5}{2}$.

$$y = \frac{1}{2} \Rightarrow \frac{x^2}{x+1} = \frac{1}{2} \Rightarrow 2x^2 - x - 1 = 0 \Rightarrow (2x+1)(x-1) = 0$$
$$\Rightarrow x_1 = -\frac{1}{2}, x_2 = 1.$$
$$y = -\frac{5}{2} \Rightarrow \frac{x^2}{x+1} = -\frac{5}{2} \Rightarrow 2x^2 + 5x + 5 = 0 \Rightarrow \text{no real solution.}$$

Thus, $x_1 = -\dfrac{1}{2}, x_2 = 1$.

6. From both sides of the given equation minus one, it is obtained that

$$\frac{8x}{3x^2 - 4x - 1} = \frac{8x}{x^2 - 4x + 1},$$
$$8x[(3x^2 - 4x - 1) - (x^2 - 4x + 1)] = 0,$$
$$x(x^2 - 1) = 0,$$
$$\text{therefore } x_1 = 0, \quad x_2 = 1, \quad x_3 = -1.$$

7. It is clear that $x = y = z = 0$ is a solution, and $x = y = z = 0$ if one of x, y, z is zero.

When $xyz \neq 0$, by taking reciprocals to both sides of the first equation, it follows that $\left(\dfrac{1}{3x}\right)^2 + 1 = \dfrac{2}{3y}$. Similarly,

$$\left(\frac{1}{3y}\right)^2 + 1 = \frac{2}{3z} \quad \text{and} \quad \left(\frac{1}{3z}\right)^2 + 1 = \frac{2}{3x}.$$

Let $u = \dfrac{1}{3x}, v = \dfrac{1}{3y}, w = \dfrac{1}{3z}$, then $u^2 - 2v + 1 = 0, v^2 - 2w + 1 = 0, w^2 - 2u + 1 = 0$. By adding them up, it is obtained that

$$(u - 1)^2 + (v - 1)^2 + (w - 1)^2 = 0,$$

so $u = v = w = 1$, hence $x = y = z = \dfrac{1}{3}$. Thus, $x = y = z = 0$ or $x = y = z = \dfrac{1}{3}$.

8. Let $x_1, x_2, \ldots, x_{100}$ be a positive solution. Since $x + y \geq 2\sqrt{xy}$ for any $x, y > 0$ and the equality holds if and only if $x = y$ if $x, y > 0$,

$$x_1 + \frac{1}{x_2} \geq 2\sqrt{\frac{x_1}{x_2}},$$
$$x_2 + \frac{1}{x_3} \geq 2\sqrt{\frac{x_2}{x_3}},$$
$$\vdots$$
$$x_{99} + \frac{1}{x_{100}} \geq 2\sqrt{\frac{x_{99}}{x_{100}}},$$
$$x_{100} + \frac{1}{x_1} \geq 2\sqrt{\frac{x_{100}}{x_1}}.$$

Multiplying all these inequalities, it is obtained that

$$4^{50} = \left(x_1 + \frac{1}{x_2}\right)\left(x_2 + \frac{1}{x_3}\right)\cdots\left(x_{100} + \frac{1}{x_1}\right) \geq 2^{100} = 4^{50},$$

therefore all the inequalities are equalities, so

$$x_1 = \frac{1}{x_2}, \quad x_2 = \frac{1}{x_3}, \cdots, x_{100} = \frac{1}{x_1}.$$

By substituting each of them into the given equations, we have

$$x_1 = x_3 = \cdots = x_{99} = 2, \quad x_2 = x_4 = \cdots = x_{100} = \frac{1}{2}.$$

9.   It is clear that the given equation has no solution when $a = b$.

When $a \neq b$, by letting $y = \dfrac{a + x}{b + x}$, the equation becomes

$$y + \frac{1}{y} = \frac{5}{2},$$
$$2y^2 - 5y + 2 = 0,$$
$$(2y - 1)(y - 2) = 0,$$

therefore $y_1 = \dfrac{1}{2}, \; y_2 = 2$.

Then

$$y = \frac{a + x}{b + x} = \frac{1}{2} \Rightarrow 2a + 2x = b + x \Rightarrow x = b - 2a.$$
$$y = \frac{a + x}{b + x} = 2 \Rightarrow a + x = 2b + 2x \Rightarrow x = a - 2b.$$

Thus, $x = a - 2b$ or $b - 2a$ when $a \neq b$, and no solution when $a = b$.

10.   From both sides of the given equation minus 2, then

$$\frac{a - x}{x + b} + \frac{a - x}{x + c} = \frac{2(a - x)}{x + b + c},$$
$$(a - x)[(2x + b + c)(x + b + c) - 2(x + b)(x + c)] = 0,$$
$$(a - x)[(b + c)x - (b^2 + c^2)] = 0,$$

therefore $x_1 = a, \; x_2 = -\dfrac{b^2 + c^2}{b + c}$.

11. From the given equation,
$$x^2 + (x + 1)^2 = 4x + a,$$
$$2x^2 - 2x + 1 - a = 0.$$

When $\Delta = 0$, then $4 - 8(1 - a) = 0$, i.e. $a = \dfrac{1}{2}$.

When $\Delta > 0$ but $x = 0$ is one solution (which is not acceptable so that there is only one real root), then $a = 1$.

When $\Delta > 0$ but $x = -1$ is one solution (which is not acceptable so that there is only one real root), then $a = 5$.

Thus $a = 1, 5$ or $\dfrac{1}{2}$.

Testing Questions   (1-B)

1. The given equation $\Leftrightarrow \dfrac{x^2 - x + 1}{x^2 + 2} + \dfrac{x^2 + 2}{x^2 - x + 1} = 2$. Let $y = \dfrac{x^2 - x + 1}{x^2 + 2}$,
   then $y + \dfrac{1}{y} = 2$ which implies $(y - 1)^2 = 1$, so $y = 1$. Thus, $x^2 - x + 1 = x^2 + 2$ gives the solution $x = -1$.

2. By long-division, the given equation is converted to the form
$$(x + 2) + \frac{x + 4}{x^2 + 5x + 13} = (x + 2) + \frac{2x + 5}{2x^2 + 7x + 20}.$$
   After canceling $x + 2$ and taking reciprocals to both sides, it follows that
$$(x + 1) + \frac{9}{x + 4} = (x + 1) + \frac{15}{2x + 5}.$$
   From $\dfrac{9}{x + 4} = \dfrac{15}{2x + 5}$ the solution $x = 5$ is obtained at once.

3. If a, b, c, d satisfy the required conditions, then so do $\dfrac{1}{a}, \dfrac{1}{b}, \dfrac{1}{c}, \dfrac{1}{d}$. Therefore
$$\frac{1}{1 - \frac{1}{a}} + \frac{1}{1 - \frac{1}{b}} + \frac{1}{1 - \frac{1}{c}} + \frac{1}{1 - \frac{1}{d}} = s.$$
   Considering that $\dfrac{1}{1 - a} + \dfrac{1}{1 - b} + \dfrac{1}{1 - c} + \dfrac{1}{1 - d} = s$ and that $\dfrac{1}{1 - x} + \dfrac{1}{1 - \frac{1}{x}} = 1$ provided $x \neq 0$ and $x \neq 1$, we have $2s = 4$, so $s = 2$.

4. It's clear that $x, y > 0$ and $y + 3x \neq 0$ for any real solution (x, y). By adding up the two equations and subtracting the first equation from the second equation respectively, the following system is obtained

$$\frac{1}{\sqrt{x}} + \frac{3}{\sqrt{y}} = 1, \tag{15.5}$$

$$-\frac{1}{\sqrt{x}} + \frac{3}{\sqrt{y}} = \frac{12}{y + 3x}. \tag{15.6}$$

By (15.5) × (15.6), then $\dfrac{9}{y} - \dfrac{1}{x} = \dfrac{12}{y + 3x}$, so

$$y^2 + 6xy - 27x^2 = 0,$$

$$(y - 3x)(y + 9x) = 0, \quad \text{i.e. } y = 3x \text{ or } y = -9x.$$

Since $x, y > 0$, so $y = -9x$ is not acceptable. By substituting $y = 3x$ into (15.5),

$$\frac{1}{\sqrt{x}} + \frac{3}{\sqrt{3x}} = 1 \Rightarrow x = (1 + \sqrt{3})^2 = 4 + 2\sqrt{3}, \ y = 12 + 6\sqrt{3}.$$

5. Since $x, y, z > 0$, the AM-GM inequality gives

$$x + \frac{1}{2x} - 1 = \frac{x}{2} + \left(\frac{x}{2} + \frac{1}{2x}\right) - 1 \geq \frac{x}{2} > 0$$

and, similarly $y + \dfrac{1}{2y} - 1 > \dfrac{y}{2} > 0, z + \dfrac{1}{2z} - 1 > \dfrac{z}{2} > 0$. Therefore

$$\left(1 - \frac{xy}{z}\right)\left(1 - \frac{yz}{x}\right)\left(1 - \frac{zx}{y}\right) > 0.$$

Among the three factors on the left hand side of the last inequality if two are negative, say $1 - \dfrac{xy}{z} < 0$ and $1 - \dfrac{yz}{x} < 0$, then $\dfrac{xy}{z} > 1, \dfrac{yz}{x} > 1$ implies $y^2 > 1$, which contradicts $0 < y < 1$. Thus,

$$1 - \frac{xy}{z} > 0, \qquad 1 - \frac{yz}{x} > 0, \qquad 1 - \frac{zx}{y} > 0.$$

On the other hand, since $0 < x, y, z < 1$, so $\dfrac{y + z}{yz} = \dfrac{1}{y} + \dfrac{1}{z} > 2$, so

$$\left(x + \frac{1}{2x} - 1\right)^2 - \left(1 - \frac{xy}{z}\right)\left(1 - \frac{zx}{y}\right)$$

$$= x^2 + 2x\left(\frac{1}{2x} - 1\right) + \left(\frac{1}{2x} - 1\right)^2 - \left[1 - \frac{x}{yz}(y^2 + z^2) + x^2\right]$$

$$= \left(\frac{1}{2x} - 1\right)^2 + \frac{x}{yz}(y - z)^2 \geq 0,$$

where the equality holds if and only if $x = \frac{1}{2}$ and $y = z$. Similarly,

$$\left(y + \frac{1}{2y} - 1\right)^2 \geq \left(1 - \frac{yz}{x}\right)\left(1 - \frac{xy}{z}\right)$$

and

$$\left(z + \frac{1}{2z} - 1\right)^2 \geq \left(1 - \frac{zx}{y}\right)\left(1 - \frac{yz}{x}\right).$$

Multiplying the three inequalities gives that the left hand side of given equation is always not less than the right hand side, and the equality holds if and only if $x = y = z = \dfrac{1}{2}$.

6. The equation $x + \dfrac{1}{y} = k$ gives $x = \dfrac{ky - 1}{y}$, i.e.

$$\frac{1}{x} = \frac{y}{ky - 1}. \tag{15.7}$$

The equation $y + \dfrac{1}{z} = k$ gives $\dfrac{1}{z} = k - y$, i.e.

$$z = \frac{1}{k - y}. \tag{15.8}$$

By substituting (15.7) and (15.8) into $z + \dfrac{1}{x} = k$, then

$$\frac{1}{k - y} + \frac{y}{ky - 1} = k. \tag{15.9}$$

From

$$\begin{aligned}
(15.9) \quad &\Leftrightarrow \quad ky - 1 + y(k - y) = k(k - y)(ky - 1) \\
&\Leftrightarrow \quad k^3 y - k^2 - k^2 y^2 + 1 - ky + y^2 = 0 \\
&\Leftrightarrow \quad ky(k^2 - 1) - (k^2 - 1) - y^2(k^2 - 1) = 0 \\
&\Leftrightarrow \quad (k^2 - 1)(ky - 1 - y^2) = 0.
\end{aligned}$$

$k^2 - 1 = 0$ gives $k = \pm 1$, and $ky - 1 - y^2 = 0$ gives $k = y + \dfrac{1}{y}$. However,

$k = y + \dfrac{1}{y}$ implies that $x = y = z$ which contradicts the assumption.

$k = \pm 1$ is possible: $x = 2, y = -1, z = \dfrac{1}{2} \Rightarrow k = 1$; and $x = -2, y = 1, z = -\dfrac{1}{2} \Rightarrow k = -1$.

7. $x, y, z > 3$ gives $y + z - 2 > 0, z + x - 4 > 0, x + y - 6 > 0$. The Cauchy-Schwartz inequality then gives

$$[y+z-2+z+x-4+x+y-6]\cdot\left[\frac{(x+2)^2}{y+z-2} + \frac{(y+4)^2}{z+x-4} + \frac{(z+6)^2}{x+y-6}\right]$$

$$\geq (x+y+z+12)^2$$

$$\Leftrightarrow \frac{(x+2)^2}{y+z-2} + \frac{(y+4)^2}{z+x-4} + \frac{(z+6)^2}{x+y-6} \geq \frac{(x+y+z+12)^2}{2(x+y+z-6)}.$$

Therefore the given equality gives the inequality

$$\frac{(x+y+z+12)^2}{x+y+z-6} \leq 72, \qquad (15.10)$$

and the equality holds if and only if $\dfrac{x+2}{y+z-2} = \dfrac{y+4}{z+x-4} = \dfrac{z+6}{x+y-6}$

$= \lambda$, namely,

$$\begin{cases} \lambda(y+z) - x = 2(\lambda+1), \\ \lambda(z+x) - y = 4(\lambda+1), \\ \lambda(x+y) - z = 6(\lambda+1). \end{cases} \qquad (15.11)$$

Let $w = x + y + z + 12$, then (15.10) becomes $\dfrac{w^2}{w-18} \leq 72$. Since

$$\frac{w^2}{w-18} \geq 72 \Leftrightarrow w^2 - 72w + 36^2 \geq 0 \Leftrightarrow (w-36)^2 \geq 0,$$

so $\dfrac{w^2}{w-18} = 72$, i.e. $w = 36$ which means $x + y + z = 24$. (15.11) gives

$$(2\lambda - 1)(x+y+z) = 12(\lambda+1),$$

by solving $2(2\lambda - 1) = \lambda + 1$ we obtain $\lambda = 1$. (15.11) then becomes

$$\begin{cases} y + z - x = 4, \\ z + x - y = 8, \\ x + y - z = 12. \end{cases}$$

By solving the system, it is found that $(x, y, z) = (10, 8, 6)$ is the unique solution.

8. (a) $a = 0 \Rightarrow x = -y$, so $2x^2 = 2 \Rightarrow x = \pm 1 \Rightarrow (x, y) = (1, 1)$ or $(-1, 1)$.

    (b) When $a \neq 0$, let $x + y = t, xy = s$. Then

$$t^2 - 2s = a^2 + 2 \qquad \text{and} \qquad \frac{t}{s} = a.$$

Therefore

$$a^2s^2 - 2s - (a^2 + 2) = 0 \Rightarrow (a^2s - a^2 - 2)(s + 1) = 0$$

$$\Rightarrow s = -1 \text{ or } s = \frac{a^2 + 2}{a^2},$$

so that $(s, t) = (-1, -a)$ or $\left(\dfrac{a^2 + 2}{a^2}, \dfrac{a^2 + 2}{a} \right)$.

(i) If $(s, t) = (-1, -a)$, then x and y are roots of the equation $z^2 + az - 1 = 0$. Since $\Delta = a^2 + 4 > 0$, it has exactly two distinct roots z_1, z_2, so (z_1, z_2) and (z_2, z_1) are two roots of the original system.

(ii) Thus, the system has no other solutions for $(s, t) = \left(\dfrac{a^2 + 2}{a^2}, \dfrac{a^2 + 2}{a} \right)$, namely, the equation

$$z^2 - \frac{a^2 + 2}{a}z + \frac{a^2 + 2}{a^2} = 0$$

has no real solution, so its discriminant $\Delta < 0$, i.e., $a^2 < 2$. Therefore $a \in (-\sqrt{2}, 0) \cup (0, \sqrt{2})$. Since $a = 0$ also let the system have exact two solutions, so the range of a is $(-\sqrt{2}, \sqrt{2})$.

9. Let $a = \dfrac{x}{z}, b = \dfrac{y}{x}, c = \dfrac{z}{y}$, then $abc = 1$ and

$$a + 1 = \frac{x}{z} + 1 = \frac{1}{zx} \cdot \frac{z}{y} = \left(\frac{z}{y} + 1 \right) \cdot \frac{z}{y} = c(c + 1) = c^2 + c.$$

Similarly, $b + 1 = a^2 + a$ and $c + 1 = b^2 + b$. Adding up these three equalities, we obtain $a^2 + b^2 + c^2 = 3$. Therefore

$$3 = a^2 + b^2 + c^2 \geq 3(abc)^{\frac{2}{3}} = 3 \Rightarrow a^2 = b^2 = c^2 = 1,$$

hence among a, b, c two are -1 and one is 1, or three are all 1.

(i) When $a = b = -1, c = 1$, then $y = z = -x$, no solution.
Similarly, no solution if $a = c = -1, b = 1$ or $b = c = -1, a = 1$.

(ii) When $a = b = c = 1$, then $x = y = z$, so that we have solutions

$$(x, y, z) = \left(\frac{1}{\sqrt{2}}, \frac{1}{\sqrt{2}}, \frac{1}{\sqrt{2}} \right) \text{ or } \left(-\frac{1}{\sqrt{2}}, -\frac{1}{\sqrt{2}}, -\frac{1}{\sqrt{2}} \right).$$

Thus, the system has real solutions

$$(x, y, z) = \left(\frac{1}{\sqrt{2}}, \frac{1}{\sqrt{2}}, \frac{1}{\sqrt{2}}\right) \text{ or } \left(-\frac{1}{\sqrt{2}}, -\frac{1}{\sqrt{2}}, -\frac{1}{\sqrt{2}}\right).$$

Solutions to Testing Questions 2

Testing Questions   (2-A)

1. Since $231 = 3 \times 7 \times 11$, letting $y = x - 2$ gives

$$(y + 3)(y + 7)(y + 11) = 231,$$
$$y^3 + 21y^2 + 131y = 0,$$
$$y(y^2 + 21y + 131) = 0,$$
$$\therefore y = 0 \ (y^2 + 21y + 131 = 0 \text{ has no real root}) \Rightarrow x = 2.$$

2. Let $y = x + 5$, then

$$(y - 3)(y - 1)(y + 1)(y + 3) = 48,$$
$$(y^2 - 9)(y^2 - 1) = 48,$$
$$y^4 - 10y^2 - 39 = 0,$$
$$(y^2 + 3)(y^2 - 13) = 0, \ \therefore y^2 = 13.$$

Thus $x_1 = -\sqrt{13} - 5, x_2 = \sqrt{13} - 5$.

3. Let $y = x - 4$, then $(y + 3)^4 + (y - 3)^4 = 272$. Since

$$(y + 3)^4 = y^4 + 12y^3 + 54y^2 + 108y + 81 \text{ and}$$
$$(y - 3)^4 = y^4 - 12y^3 + 54y^2 - 108y + 81,$$
$$y^4 + 54y^2 + 81 - 136 = 0,$$
$$(y^2 + 55)(y^2 - 1) = 0,$$
$$\therefore y^2 = 1, \ \text{i.e. } y = \pm 1.$$

Thus, $x_1 = 3, x_2 = 5$.

4. Since $x \neq 0$, so

$$2(x^2 + 1) + 7(x^3 + x) + 6x^2 = 0,$$
$$2\left(x^2 + \frac{1}{x^2}\right) + 7\left(x + \frac{1}{x}\right) + 6 = 0.$$

Let $y = x + \dfrac{1}{x}$, then

$$2y^2 + 7y + 6 = 0,$$
$$(2y + 3)(y + 2) = 0, \quad \therefore y = -\frac{3}{2} \text{ or } y = -2.$$

Thus,

$$y = -\frac{3}{2} \Rightarrow x + \frac{1}{x} = -\frac{3}{2} \Rightarrow 2x^2 + 3x + 2 = 0, \text{ no real solution.}$$

$$y = -2 \Rightarrow x + \frac{1}{x} = -2 \Rightarrow x^2 + 2x + 1 = 0 \Rightarrow x_1 = x_2 = -1.$$

5. By using the substitution $y = x^2$, the quadratic equation $y^2 - (k-1)y + (2-k) = 0$ is obtained, and it has two positive different real roots, therefore its discriminant is positive.

$$(k-1)^2 - 4(2-k) > 0, \quad \text{and} \quad \frac{(k-1) - \sqrt{(k-1)^2 - 4(2-k)}}{2} > 0.$$

$k^2 + 2k - 7 > 0$ gives that $k < -1 - 2\sqrt{2}$ or $k > -1 + 2\sqrt{2}$.

$\dfrac{(k-1) - \sqrt{(k-1)^2 - 4(2-k)}}{2} > 0$ implies that $k - 1 > 0$ and $2 - k > 0$,

i.e. $1 < k < 2$, therefore the range of k is

$$-1 + 2\sqrt{2} < k < 2.$$

6. Since

$$\alpha^3 - 3\alpha^2 + 5\alpha - 17 = (\alpha^3 - 3\alpha^2 + 3\alpha - 1) + 2(\alpha - 1) - 14$$

and

$$-(\beta^3 - 3\beta^2 + 5\beta + 11) = (1 - 3\beta + 3\beta^2 - \beta^3) + 2(1 - \beta) - 14,$$

each of $\alpha - 1$ and $1 - \beta$ is the real root of equation $x^3 + 2x - 14 = 0$. Since the cubic equation has only one real root, so

$$\alpha - 1 = 1 - \beta, \quad \therefore \alpha + \beta = 2.$$

7. Considering that the variables are cyclic in the problem, without loss of generality we may assume that $z \geq x, y$.

If $x > y$, then $5z^3 = y^5 + 4x < x^5 + 4z = 5y^3$, a contradiction.

If $x \leq y$, then $5z^3 = y^5 + 4x \leq z^5 + 4y = 5x^3$, so $x = y = z$.

Thus, $x^5 - 5x^3 + 4x = x(x^2 - 1)(x^2 - 4) = 0$, i.e., $x = 0, \pm 1$ or $x = \pm 2$, there are 5 solutions for (x, y, z) in total.

8. Let

$$x^3 + y^3 = 7, \tag{15.12}$$
$$xy(x + y) = -2. \tag{15.13}$$

By (15.12) + 3 × (15.13), we obtain

$$(x + y)^3 = 1.$$

Therefore $x + y = 1$ and $xy = -2$. So (x, y) are roots of the equation $t^2 - t - 2 = 0$.

Thus, $(x, y) = (2, -1)$ or $(-1, 2)$.

9. Since $x^3 + y^3 = (x + y)(x^2 - xy + y^2)$, all pairs of integers $(n, -n), n \in \mathbb{Z}$, are solutions.

Suppose that $x + y \neq 0$. Then the equation becomes

$$x^2 - xy + y^2 = x + y,$$

i.e.

$$x^2 - (y + 1)x + y^2 - y = 0.$$

Treated as a quadratic equation in x, we calculate the discriminant

$$\Delta = y^2 + 2y + 1 - 4y^2 + 4y = -3y^2 + 6y + 1.$$

Solving for $\Delta \geq 0$ yields

$$\frac{3 - 2\sqrt{3}}{3} \leq y \leq \frac{3 + 2\sqrt{3}}{3}.$$

Thus the possible values for y are 0, 1, and 2, which lead to the solutions $(1, 0), (0, 1), (1, 2), (2, 1)$, and $(2, 2)$.

Therefore, the integer solutions of the equation are

$$(x, y) = (1, 0), (0, 1), (1, 2), (2, 1), (2, 2) \text{ and } (n, -n), \text{ for all } n \in \mathbb{Z}.$$

10. Multiplying the n given equations, then

$$\prod_{i=1}^{n} x_i^2(3a - 2x_i) = a^{3n}. \tag{15.14}$$

Since $a > 0$ and $x_i > 0, i = 1, 2, \ldots, n$, by the AM-GM inequality,

$$x_i^2(3a - 2x_i) \leq \left(\frac{x_i + x_i + 3a - 2x_i}{3} \right)^3 = a^3$$

for $i = 1, 2, \ldots, n$, so $x_i = 3a - 2x_i$, i.e., $x_i = a, i = 1, 2, \ldots, n$.

Thus, the unique solution of the given equation is $(a, a, \ldots, a)$.

11. The second equation minus the first equation gives

$$(x^3 - y^3) - (x^2 - y^2) + (x - y) = 0,$$

so $(x - y)(x^2 + xy + y^2 - x - y + 1) = 0$. Since

$$x^2 + xy + y^2 - x - y + 1 = \frac{1}{2}[(x + y)^2 + (x - 1)^2 + (y - 1)^2] > 0,$$

it is obtained that $x = y$. Hence the given equation becomes $x + x^2 = x^3$, and its solution are 0, $\frac{1}{2}(1 - \sqrt{5})$ and $\frac{1}{2}(1 + \sqrt{5})$. Thus, the solutions of the original system are

$$(0, 0), \quad \left(\frac{1}{2}(1 - \sqrt{5}), \frac{1}{2}(1 - \sqrt{5})\right), \quad \left(\frac{1}{2}(1 + \sqrt{5}), \frac{1}{2}(1 + \sqrt{5})\right).$$

Testing Questions   (2-B)

1. It is obvious that $(0, 0, 0)$ is a solution, and it is easy to show that if one of x, y, z is zero then the other two are zeros also.

Below we assume that $xyz \neq 0$.

Let $a = \dfrac{x}{y}, b = \dfrac{z}{y}$. Then the system becomes

$$a + 1 + b = 3ay, \quad a^2 + 1 + b^2 = 3ab, \quad y(a^3 + 1 + b^3) = 3b.$$

Use $\dfrac{1 + a + b}{3a}$ to replace y in the system gives

$$1 + a^2 + b^2 = 3ab \quad \text{and} \quad (1 + a + b)(1 + a^3 + b^3) = 9ab.$$

Let $u = a + b$ and $v = ab$, then

$$1 + u^2 - 2v = 3v \quad \text{and} \quad (1 + u)(1 + u^3 - 3uv) = 9v.$$

From the first equation we have $v = \frac{u^2+1}{5}$, so from the second equation

$$u^4 + u^3 - 6u^2 + u - 2 = 0,$$
$$(u - 2)(u^3 + 3u^2 + 1) = 0.$$

$u = 2$ implies $v = 1$, so that $a = b = 1$, which means $x = y = z = 1$. Thus, we obtain a solution $(1, 1, 1)$.

Since $f(u) = u^3 + 3u^2 + 1$ has a local maximum value 5 at $u = -2$ and a local minimum value 1 at $u = 0$, so $f(u) = 0$ has unique real root u_0 at $(-\infty, -2)$. By Viete's Theorem, a, b are the roots of the equation $t^2 - u_0 t + \frac{u_0^2 + 1}{5} = 0$. Since

$$\Delta = u_0^2 - 4\left(\frac{u_0^2 + 1}{5}\right) = \frac{u_0^2 - 4}{5} > 0,$$

the equation has two distinct real solutions for (a, b). By exchanging a and b, each pair $\{a, b\}$ gives two solutions for (x, y, z), so there are 6 solutions for (x, y, z) in total.

2. Substituting $y = kx + d$ into $x^3 + y^3 = 2$, then

$$(k^3 + 1)x^3 + 3k^2 dx^2 + 3kd^2 x + d^3 - 2 = 0.$$

Since every cubic equation must have at least one real root, by substituting the x into above linear equation, we can get a corresponding y. Therefore $k^3 + 1 = 0$ so that the system has no real root. Thus, the cubic equation becomes

$$3dx^2 - 3d^2 x + d^3 - 2 = 0.$$

When $d = 0$, then the equation becomes $-2 = 0$, i.e. no solution.

When $d \neq 0$, the quadratic equation has no real roots if and only if its discriminant is negative, i.e. $9d^4 - 12d(d^3 - 2) < 0$. Since
$9d^4 - 12d(d^3 - 2) < 0 \Leftrightarrow 9d^4 - 12d^4 + 24d < 0 \Leftrightarrow 24d - 3d^4 < 0$
$\Leftrightarrow 3d(2^3 - d^3) < 0 \Leftrightarrow 3d(2 - d)(4 + 2d + d^2) < 0$
$\Leftrightarrow d(2 - d)[(d + 1)^2 + 3] < 0 \Leftrightarrow d < 0$ or $d > 2$.

Thus, the conditions on k and d are $k = -1, d \leq 0$ or $d > 2$.

3. $(2.3) - (2.4)$ yields

$$x^3 - y^3 + x^2 y - xy^2 + x^2 - y^2 = 0,$$
$$(x - y)(x^2 + xy + y^2 + xy + x + y) = 0,$$
$$(x - y)(x + y)(x + y + 1) = 0.$$

When $x = y$, then Eq.(2.3) gives

$$x^3 + 1 - x^3 - x^2 = 0 \Rightarrow x^2 = 1 \Rightarrow (x, y) = (1, 1) \text{ or } (-1, -1).$$

When $x + y = 0$, then Eq.(2.3) gives

$$x^3 + 1 - x^3 - x^2 = 0 \Rightarrow x^2 = 1 \Rightarrow (x, y) = (1, -1) \text{ or } (-1, 1).$$

When $x + y + 1 = 0$, then Eq.(2.3) gives

$$x^3 + 1 - x(x+1)^2 - (x+1)^2 = 0 \Rightarrow x^3 + 1 - (x+1)^3 = 0$$
$$\Rightarrow 3x(x+1) = 0 \Rightarrow x = 0 \text{ or } -1 \Rightarrow (x, y) = (0, -1) \text{ or } (-1, 0).$$

In summary, $(x, y) = (1, 1), (-1, -1), (1, -1), (-1, 1), (0, -1), (-1, 0)$.

4. Consider the given equation as a quadratic equation in a:

$$a^2 + 3xa + 2x^2 - x^3 - x^4 = 0.$$

The discriminant of this equation is

$$9x^2 - 8x^2 + 4x^3 + 4x^4 = (x + 2x^2)^2.$$

Thus

$$a = \frac{-3x + (x + 2x^2)}{2} = -x + x^2 \qquad \text{or} \qquad a = -2x - x^2.$$

The first choice $a = -x + x^2$ yields the quadratic equation $x^2 - x - a = 0$, whose solutions are $x = \dfrac{1 \pm \sqrt{1 + 4a}}{2}$.

The second choice $a = -2x - x^2$ yields the quadratic equation $x^2 + 2x + a = 0$, whose solutions are

$$x = -1 \pm \sqrt{1 - a}.$$

The inequalities

$$-1 - \sqrt{1 - a} < -1 + \sqrt{1 - a} < \frac{1 - \sqrt{1 + 4a}}{2} < \frac{1 + \sqrt{1 + 4a}}{2}$$

show that the four solutions are distinct. Indeed,

$$-1 + \sqrt{1 - a} < \frac{1 - \sqrt{1 + 4a}}{2} \Leftrightarrow 2\sqrt{1 - a} < 3 - \sqrt{1 + 4a}$$
$$\Leftrightarrow 6\sqrt{1 + 4a} < 6 + 8a \Leftrightarrow 3a < 4a^2 \Leftrightarrow \frac{3}{4} < a.$$

5. Multiplying the first equation by y, the second by x, and adding up yields

$$2xy + \frac{(3x - y)y - (x + 3y)x}{x^2 + y^2} = 3y,$$

or $2xy - 1 = 3y$. It follows that $y \neq 0$ and $x = \dfrac{3y + 1}{2y}$.

Substituting this into the second equation of the given system gives

$$y\left[\left(\frac{3y+1}{2y}\right)^2 + y^2\right] - \left(\frac{3y+1}{2y}\right) - 3y = 0,$$

or

$$4y^4 - 3y^2 - 1 = 0.$$

It follows that $y^2 = 1$ and that the solutions to the system are $(2, 1)$ and $(1, -1)$.

6. Let $S = \dfrac{r-1}{r+1} + \dfrac{s-1}{s+1} + \dfrac{t-1}{t+1}$ and $R = \dfrac{1}{r+1} + \dfrac{1}{s+1} + \dfrac{1}{t+1}$, then

$$S = 3 - 2R.$$

r, s, t are roots of the $P(x)$ implies that $r + 1, s + 1, t + 1$ are roots of $Q(x)$ defined by

$$Q(x) = P(x - 1) = x^3 - 3x^2 - 2004x + 4008,$$

by Viete's Theorem,

$$R = \frac{(s+1)(t+1) + (r+1)(t+1) + (r+1)(s+1)}{(r+1)(s+1)(t+1)} = \frac{-2004}{-4008} = \frac{1}{2}.$$

Thus, $S = 3 - 2 \cdot \dfrac{1}{2} = 2$.

7. Let

$$x + y + z = 2, \quad (15.15)$$
$$(x + y)(y + z) + (y + z)(z + x) + (z + x)(x + y) = 1, \quad (15.16)$$
$$x^2(y + z) + y^2(z + x) + z^2(x + y) = -6. \quad (15.17)$$

(15.16) implies that $x^2 + y^2 + z^2 + 3xy + 3yz + 3zx = 1$, i.e.

$$(x + y + z)^2 + xy + yz + zx = 1.$$

Combining with (15.15), it follows that

$$xy + yz + zx = -3. \quad (15.18)$$

Substituting (15.15), (15.18) into (15.17), then

$$x(xy + xz) + y(yz + xy) + z(xz + yz) = -6$$
$$\Rightarrow x(3 + yz) + y(3 + xz) + z(3 + xy) = 6$$
$$\Rightarrow x + y + z + xyz = 2,$$

hence

$$xyz = 0. \tag{15.19}$$

By Viete's theorem, (15.15), (15.18) and (15.19) implies that x, y, z are the three roots of the equation $t^3 - 2t^2 - 3t = 0$, which has three roots $0, -1$ and 3.

Thus, the roots (x, y, z) of the original equation are

$$(0, 3, -1), \ (0, -1, 3), \ (3, 0, -1), \ (3, -1, 0), \ (-1, 0, 3), \ (-1, 3, 0).$$

8. From the given equations, $a, b, c, d, t > 0$ implies $0 < a, b, c, d < 1$. If two of them are not equal, then there must be consecutive two not equal. Without loss of generality, we may assume $a \neq b$.

    If $a < b$, then $a(1 - b^2) = b(1 - c^2) \Rightarrow b < c$, and similarly, $b < c \Rightarrow c < d \Rightarrow d < a$, so $a < b < c < d < a$, a contradiction.

    A contradiction is obtainable if $a > b$, therefore $a = b = c = d$. Thus it suffices to find solutions of

    $$a(1 - a^2) = t. \tag{15.20}$$

    $a > 1$ implies $a(1 - a^2) < 0$ and $a = 0, 1$ implies $a(1 - a^2) = 0$, so $0 < a < 1$. Let $M = \max\limits_{0 < a < 1} \{a(1 - a^2)\}$. Then (15.20) has no solution if $t > M$; has one solution if $t = M$; and two solutions if $0 < t < M$.

    The value of M can be obtained as follows: Suppose that the equation $a(1 - a^2) = M$ (i.e. $a^3 - a + M = 0$) has real roots $\{\alpha, \alpha, \beta\}$, where $\alpha > 0, \beta < 0$. Then $2\alpha + \beta = 0, \alpha^2 + 2\alpha\beta = -1$. By solving them it is obtained that $\alpha = 1/\sqrt{3}, \beta = -2/\sqrt{3}$, hence

    $$M = \frac{1}{\sqrt{3}}\left(1 - \frac{1}{3}\right) = \frac{2\sqrt{3}}{9}.$$

    Thus, the original equation has no solution if $t > \dfrac{2\sqrt{3}}{9}$; has one solution if $t = \dfrac{2\sqrt{3}}{9}$; and two solutions if $0 < t < \dfrac{2\sqrt{3}}{9}$.

9. We consider the following cases.

    1. $xy = 0$. Then it is clear that $x = y = 0$ and $(x, y) = (0, 0)$ is a solution.

    2. $xy < 0$. By the symmetry, we can assume that $x > 0 > y$. Then $(1 + x)(1 + x^2)(1 + x^4) > 1$ and $1 + y^7 < 1$. There are no solutions in this case.

3. $x, y > 0$ and $x \neq y$. By the symmetry, we can assume that $x > y > 0$. Then

$$(1 + x)(1 + x^2)(1 + x^4) > 1 + x^7 > 1 + y^7,$$

showing that there are no solutions in this case.

4. $x, y < 0$ and $x \neq y$. By the symmetry, we can assume that $x < y < 0$. Multiplying by $1 - x$ and $1 - y$ the first and the second equation, respectively, the system now reads

$$1 - x^8 = (1 + y^7)(1 - x) = 1 - x + y^7 - xy^7,$$
$$1 - y^7 = (1 + x^7)(1 - y) = 1 - y + x^7 - x^7 y.$$

Subtracting the first equation from the second yields

$$x^8 - y^8 = (x - y) + (x^7 - y^7) - xy(x^6 - y^6). \tag{15.21}$$

Since $x < y < 0, x^8 - y^8 > 0, x - y < 0, x^7 - y^7 < 0, -xy < 0$, and $x^6 - y^6 > 0$. Therefore, the left-hand side of (15.21) is positive while the right-hand side of (15.21) is negative.
Thus there are no solutions in this case.

5. $x = y$. Then solving

$$1 - x^8 = 1 - x + y^7 - xy^7 = 1 - x + x^7 - x^8$$

leads to $x = 0, 1, -1$, which implies that $(x, y) = (0, 0)$ or $(-1, -1)$.

Thus, $(x, y) = (0, 0)$ and $(-1, -1)$ are the only solutions to the system.

10. First of all, we show an inequality:

$$2x^k \geq [(k - 1)x - (k - 2)](x^2 + 1) \qquad \text{for all } k \geq 3, x \geq 0. \tag{15.22}$$

Proof: By AM-GM inequality,

$$x^k + x^k + \underbrace{x + x + \cdots + x}_{k-3} \geq (k - 1) \sqrt[k-1]{x^{3(k-1)}} = (k - 1)x^3,$$

so $2x^k \geq (k - 1)x^3 - (k - 3)x$. Since

$$(k-1)x^3 - (k-3)x - [(k-1)x - (k-2)](x^2+1) = (k-2)(x^2-2x+1) \geq 0,$$

(15.22) is proven. The proof indicates that the equality holds if and only if $x = 1$ ($\because x \geq 0$). Applying (15.22) to each of the three given equations,

then

$$\begin{cases} 2y(x^2 + 1) - (z^2 + 1) \geq (2x - 1)(x^2 + 1), \\ 3z(y^2 + 1) - 2(x^2 + 1) \geq (3y - 2)(y^2 + 1), \\ 4x(z^2 + 1) - 3(y^2 + 1) \geq (4z - 3)(z^2 + 1), \end{cases}$$

namely

$$2(y - x)(x^2 + 1) + (x - z)(x + z) \geq 0, \tag{15.23}$$

$$3(z - y)(y^2 + 1) + 2(y - x)(y + x) \geq 0, \tag{15.24}$$

$$4(x - z)(z^2 + 1) + 3(z - y)(z + y) \geq 0. \tag{15.25}$$

(i) When $x \geq \max\{y, z\}$, (15.24) implies $y \leq z$, so

$$\begin{aligned} & 2(y - x)(x^2 + 1) + (x - z)(x + z) \\ & \leq (z - x)[2(x^2 + 1) - (x + z)] \\ & \leq (z - x)(2x^2 - 2x + 2) \leq 0, \end{aligned}$$

hence $2(y - x)(x^2 + 1) + (x - z)(x + z) = 0$. Then

$$x - y \geq x - z, 2(x^2 + 1) > 2x \geq x + z \Rightarrow x = y, x = z,$$

so $x = y = z$.

(ii) When $y \geq \max\{y, z\}$, (15.25) implies $z \leq x$, so

$$\begin{aligned} & 3(z - y)(y^2 + 1) + 2(y - x)(y + x) \\ & \leq (x - y)[3(y^2 + 1) - 2y - 2x] \\ & \leq (x - y)(3y^2 - 4y + 3) \leq 0, \end{aligned}$$

hence $3(z - y)(y^2 + 1) + 2(y - x)(y + x) = 0$. Then

$$y - z \geq y - x, 3(y^2 + 1) > 2(y^2 + 1) \geq 4y \geq 2(y + x)$$

implies that $x = y$ and $y = z$, so $x = y = z$.

(iii) When $z \geq \max\{x, y\}$, (15.23) implies $x \leq y$, so

$$\begin{aligned} & 4(x - z)(z^2 + 1) + 3(z - y)(z + y) \\ & \leq (y - z)(4z^2 - 6z + 4) \leq 0, \end{aligned}$$

hence $4(x - z)(z^2 + 1) + 3(z - y)(z + y) = 0$. Then

$$z - x \geq z - y, 4(z^2 + 1) > 3(z^2 + 1) \geq 3(2z) \geq 3(z + y)$$

implies that $z = x, z = y$, so $x = y = z$.

Thus, $x = y = z$, and, by solving $2x^3 = 2x(x^2 + 1) - (x^2 + 1)$ i.e. $(x - 1)^2 = 0$, we have the unique solution $x = y = z = 1$.

Solutions to Testing Questions   3

Testing Questions   (3-A)

1. Write the equation in the form $\sqrt{7x^2 + 9x + 13} = 7x - \sqrt{7x^2 - 5x + 13}$, and taking squares to both sides, then

$$7x^2 + 9x + 13 = 49x^2 + 7x^2 - 5x + 13 - 14x\sqrt{7x^2 - 5x + 13},$$
$$2\sqrt{7x^2 - 5x + 13} = 7x - 2,$$
$$4(7x^2 - 5x + 13) = 49x^2 - 28x + 4,$$
$$21x^2 - 8x - 48 = 0 \Rightarrow (7x - 12)(3x + 4) = 0,$$
$$\therefore x = \frac{12}{7} \ \left(x = -\frac{4}{3} \text{ is not acceptable since } 7x > 0 \right).$$

2. By completing square, the given equation can be written in the form

$$x^2 + (2x)(2\sqrt{x + 3}) + (2\sqrt{x + 3})^2 - 12 = 13,$$
$$(x + 2\sqrt{x + 3})^2 = 25,$$
$$\therefore x + 2\sqrt{x + 3} = 5 \quad \text{or} \quad x + 2\sqrt{x + 3} = -5.$$

$x + 2\sqrt{x + 3} = 5 \Rightarrow x^2 - 14x + 13 = 0 \Rightarrow x_1 = 1, x_2 = 13$ (N.A.).

$x + 2\sqrt{x + 3} = -5 \Rightarrow x^2 + 6x + 13 = 0$, no real solution.

Thus, $x = 1$ is the unique real solution.

3. 0 is clearly not a solution, so the given equation can be simplified in the form

$$x^2 - 2x - 4 + 3\sqrt{x^2 - 2x} = 0.$$

Let $w = \sqrt{x^2 - 2x}$, then $w^2 + 3w - 4 = 0$, so $w = 1$. ($w = -4$ is not acceptable since $w \geq 0$). Thus,

$$w = 1 \Rightarrow \sqrt{x^2 - 2x} = 1 \Rightarrow x^2 - 2x - 1 = 0 \Rightarrow x = 1 \pm \sqrt{2}.$$

By checking, the two values both satisfy the given equation, so they are the solutions.

4. Change the given equation to the form

$$(x^2 + 1 - 4\sqrt{x^2 + 1} + 4) + (y^2 - 4 - 4\sqrt{y^2 - 4} + 4)$$
$$+ (z^2 - 1 - 2\sqrt{z^2 - 1} + 1) = 0,$$

then

$$(\sqrt{x^2+1}-2)^2 + (\sqrt{y^2-4}-2)^2 + (\sqrt{z^2-1}-1)^2 = 0,$$

so $\sqrt{x^2+1} = 2,\ \sqrt{y^2-4} = 2,\ \sqrt{z^2-1} = 1$, i.e.,

$$x = \pm\sqrt{3},\ \ y = \pm2\sqrt{2},\ \ z = \pm\sqrt{2},$$

the number of real roots (x, y, z) is 8.

5. By rationalizing the denominator of the left hand side of the equation,

$$\frac{(a+x)+(a-x)+2\sqrt{(a+x)(a-x)}}{2x} = \frac{a}{x},$$

$$\sqrt{(a+x)(a-x)} = 0,$$

$$\therefore x = \pm a.$$

By checking, the two values are both roots.

6. From the given equation,

$$|x-3| + |x+4| = 7.$$

(i)   When $x \le -4$, then $3 - x - 4 - x = 7$, therefore $x = -4$.

(ii) When $-4 < x \le 3$, then $3 - x + x + 4 = 7 \Rightarrow 0 \cdot x = 0$, solution set is the interval $-4 < x \le 3$.

(iii) When $3 < x$, then $x - 3 + x + 4 = 7$, i.e. $x = 3$, so no solution.

By checking, $-4 \le x \le 3$ is the solution set.

7. Since $x \ge \dfrac{1}{2}$, the left hand side is

$$\sqrt{x+\sqrt{2x-1}} + \sqrt{x-\sqrt{2x-1}} = \sqrt{x+\sqrt{2x-1}} + \frac{|x-1|}{\sqrt{x+\sqrt{2x-1}}}.$$

By taking squares to both side, it is obtained that

$$2x + 2|x-1| = a.$$

On the interval $x > 1$, the equation has solution $x = \dfrac{2+a}{4}$ if $a \ge 2$. On the interval $x \le 1$, the equation is $0 \cdot x + 2 = a$, so any x on $[\frac{1}{2}, 1]$ is a solution if $a = 2$. When $a < 2$ then no solution for x.

8. Let $y = \sqrt{x^2 + x + 7}$, then $y^2 + y - 12 = 0$, so $y = 3$ or -4 (N.A.).
   $y = 3 \Rightarrow \sqrt{x^2 + x + 7} = 3 \Rightarrow x^2 + x - 2 = 0 \Rightarrow x = -2$ or 1.

9. The given equation yields $\sqrt{2x + 1} - 3 = \sqrt{x + 7} - \sqrt{x + 3}$, so

$$2x + 1 + 9 - 6\sqrt{2x + 1} = 2x + 10 - 2\sqrt{x^2 + 10x + 21},$$
$$x^2 + 10x + 21 = 9(2x + 1),$$
$$x^2 - 8x + 12 = 0, \quad \therefore x = 2 \text{ or } 6.$$

Since $\sqrt{2x + 1} - 3 = \sqrt{x + 7} - \sqrt{x + 3} > 0$ implies that $x > 4$, so $x = 6$ is the unique solution.

10. The given system gives the conditions $x, y > 0$ and $y + 3y \neq 0$, and the system is equivalent to

$$\frac{1}{\sqrt{x}} + \frac{3}{\sqrt{y}} = 1, \tag{15.26}$$

$$-\frac{1}{\sqrt{x}} + \frac{3}{\sqrt{y}} = \frac{12}{y + 3y}. \tag{15.27}$$

(15.26) $\times$ (15.27) gives $\dfrac{9}{y} - \dfrac{1}{x} = \dfrac{12}{y + 3x}$

$\Leftrightarrow y^2 + 6xy - 27x^2 = 0 \Leftrightarrow y = 3x$ or $y = -9x$ (not acceptable).

From (15.26), $y = 3x \Rightarrow \dfrac{1}{\sqrt{x}} + \dfrac{3}{\sqrt{3x}} = 1 \Rightarrow x = 4 + 2\sqrt{3}, y = 12 + 6\sqrt{3}$.

11. It is easy to see that $x = 0$ is a solution. Since the right hand side is a decreasing function of x and the left hand side is an increasing function of x, there is at most one solution.

Thus $x = 0$ is the only solution to the equation.

Testing Questions   (3-B)

1. Let $y = \sqrt[3]{x + 9}$, then $x + 1 = y^3 - 8 = (y - 2)(y^2 + 4y + 4)$, and the given equation becomes

$$y - 2 = -\sqrt{(y - 2)(y^2 + 4y + 4)},$$
$$(y - 2)^2 = (y - 2)(y^2 + 4y + 4),$$
$$(y - 2)(y^2 + 3y + 6) = 0 \Rightarrow y = 2.$$

From $\sqrt[3]{x + 9} = 2$ it is easy to find that $x = -1$.

2. $\dfrac{5x - 6 - x^2}{2} \geq 0$ implies $x^2 - 5x + 6 \leq 0$, so the real solution x of the given equation must satisfy $2 \leq x \leq 3$.

Below we consider the left hand side of the given equation. Since $2 \leq x \leq 3$,

$$\sqrt{4 - x\sqrt{4 - (x - 2)\sqrt{1 + (x - 5)(x - 7)}}}$$
$$= \sqrt{4 - x\sqrt{4 - (x - 2)(6 - x)}} = \sqrt{4 - x(4 - x)} = \sqrt{(x - 2)^2} = x - 2.$$

Therefore the equation is simplified as $x - 2 = \dfrac{5x - 6 - x^2}{2}$, i.e. $x^2 - 3x + 2 = 0$. Then $(x - 1)(x - 2) = 0$ gives that

$$x = 2$$

is the unique real solution.

3. Let $u = x + y$ and $v = x - y$. Then

$$0 \leq x^2 - y^2 = uv < 1, \quad x = \frac{u + v}{2}, \quad \text{and} \quad y = \frac{u - v}{2}.$$

Adding the two given equations and subtracting the two given equations yields the new system

$$u - u\sqrt{uv} = (a + b)\sqrt{1 - uv}$$
$$v + v\sqrt{uv} = (a - b)\sqrt{1 - uv}.$$

Multiplying the above two equations yields

$$uv(1 - uv) = (a^2 - b^2)(1 - uv),$$

hence $uv = a^2 - b^2$. It follows that

$$u = \frac{(a + b)\sqrt{1 - a^2 + b^2}}{1 - \sqrt{a^2 - b^2}} \quad \text{and} \quad v = \frac{(a - b)\sqrt{1 - a^2 + b^2}}{1 + \sqrt{a^2 - b^2}},$$

which in turn implies that

$$(x, y) = \left(\frac{a + b\sqrt{a^2 - b^2}}{\sqrt{1 - a^2 + b^2}}, \frac{b + a\sqrt{a^2 - b^2}}{\sqrt{1 - a^2 + b^2}} \right),$$

whenever $0 \leq a^2 - b^2 < 1$.

4. Let $u = \sqrt[4]{x - y}$, the second equation then becomes $u^4 + 13u - 42 = 0$, i.e., $(u - 2)(u^3 + 2u^2 + 4u + 21) = 0$ which has a unique positive root $u = 2$. Therefore

$$x = y + 16.$$

Substituting it into the first equation, then

$$y^2 - 5y + 3 = 0,$$

so $y = \dfrac{5 + \sqrt{13}}{2}$ (since $\dfrac{5 - \sqrt{13}}{2} < 2$ is not acceptable).

Thus, $x = \dfrac{37 + \sqrt{13}}{2}, y = \dfrac{5 + \sqrt{13}}{2}.$

5. The question gives the conditions: $1 + 2xy > 0$, $x(1 - 2x) \geq 0$ and $y(1 - 2y) \geq 0$. By solving them, it is obtained that

$$0 \leq x \leq \frac{1}{2} \quad \text{and} \quad 0 \leq y \leq \frac{1}{2}.$$

Below we show that for $0 \leq x, y \leq \dfrac{1}{2}$

$$\frac{1}{\sqrt{1 + 2x^2}} + \frac{1}{\sqrt{1 + 2y^2}} \leq \frac{2}{\sqrt{1 + 2xy}}. \tag{15.28}$$

In fact,

$$(15.28) \Leftrightarrow \frac{1}{1 + 2x^2} + \frac{1}{1 + 2y^2} + \frac{2}{\sqrt{(1 + 2x^2)(1 + 2y^2)}} \leq \frac{4}{1 + 2xy}$$

$$\Leftrightarrow \left(\frac{1}{1 + 2x^2} + \frac{1}{1 + 2y^2} - \frac{2}{1 + 2xy} \right)$$
$$+ \left(\frac{2}{\sqrt{4x^2 y^2 + 2x^2 + 2y^2 + 1}} - \frac{2}{2xy + 1} \right) \leq 0$$

$$\Leftrightarrow \left[-\frac{2x(x - y)}{(1 + 2x^2)(1 + 2xy)} + \frac{2y(x - y)}{(1 + 2y^2)(1 + 2xy)} \right] +$$
$$\left(\frac{2}{\sqrt{4x^2 y^2 + 2x^2 + 2y^2 + 1}} - \frac{2}{2xy + 1} \right) \leq 0$$

$$\Leftrightarrow -\frac{2(x - y)^2 (1 - 2xy)}{(1 + 2xy)(1 + 2x^2)(1 + 2y^2)}$$
$$+ \left(\frac{2}{\sqrt{4x^2 y^2 + 2x^2 + 2y^2 + 1}} - \frac{2}{2xy + 1} \right) \leq 0.$$

On the left hand side, the first term is clearly less than or equal to zero, and for the second term,

$$\frac{2}{\sqrt{4x^2y^2 + 2x^2 + 2y^2 + 1}} \leq \frac{2}{\sqrt{4x^2y^2 + 4xy + 1}} = \frac{2}{2xy + 1}$$

implies that it is also less than or equal to zero, so (15.28) is proven, and the equality holds if and only if $x = y$.

Thus, the first given equation means $x = y$, and the second equation becomes

$$\sqrt{x(1 - 2x)} = \sqrt{y(1 - 2y)} = \frac{1}{9},$$

from which $x = y = \dfrac{9 \pm \sqrt{73}}{36}$.

6. Let $u = \sqrt{3x^2 + x - 1}, v = \sqrt{x^2 - 2x - 3}, w = \sqrt{3x^2 + 3x + 5}$ and $z = \sqrt{x^2 + 3}$, then

$$u + v = w + z \qquad \text{and} \qquad u^2 - v^2 = w^2 - z^2,$$

which yields $u - v = w - z$. Therefore $u = w$. Then

$$\sqrt{3x^2 + x - 1} = \sqrt{3x^2 + 3x + 5} \Rightarrow x - 1 = 3x + 5 \Rightarrow x = -3.$$

7. Since $x^2 - 4|x| + 5 = (|x| - 2)^2 + 1 > 0$, the given equation has the form

$$x^2 - 4|x| + 5 = |x - 4| + 1.$$

(i) $x \leq 0 \Rightarrow x^2 + 5x = 0 \Rightarrow x = 0$ or -5;

(ii) $0 < x \leq 4 \Rightarrow x^2 - 3x = 0 \Rightarrow x = 3$;

(iii) $4 < x \Rightarrow x^2 - 5x + 8 = 0$, no real solution.

Thus, the number of real solutions is 3.

8. Let $u = 4x^{100}, v = y^{100}$, then the given equation becomes

$$(u^2 + 1)(v^2 + 1) = 4uv.$$

From

$(u^2 + 1)(v^2 + 1) = 4uv \Leftrightarrow [(uv)^2 - 2uv + 1] + (u^2 + v^2 - 2uv) = 0$
$\Leftrightarrow uv = 1$ and $u = v \Rightarrow u = v = 1$ ($\because u, v \geq 0$),

it follows that $x = \pm \dfrac{1}{\sqrt[50]{2}}, \quad y = \pm 1$, therefore

$$(x, y) = \left(\frac{1}{\sqrt[50]{2}}, 1 \right), \left(\frac{1}{\sqrt[50]{2}}, -1 \right), \left(-\frac{1}{\sqrt[50]{2}}, 1 \right), \left(-\frac{1}{\sqrt[50]{2}}, -1 \right).$$

Solutions to Test questions 4

Testing Questions (4-A)

1. $2^{300} = 8^{100}, 3^{200} = 9^{100}$, so (A) < (B).

    $4^{100} = 2^{200} < 2^{300}$, so (C) < (A).

    $2^{100} + 3^{100} < 2 \cdot 3^{100} < 3^{101} < 3^{200}$, so (D) < (B).

    $3^{50} + 4^{50} = 3^{50} + 2^{100}$, so (E) < (D) < (B). Thus, (B) is the greatest.

2. Let $a = 15^{2010}$. Then it suffices to compare the sizes of $A = \dfrac{a+1}{15a+1}$ and

    $B = \dfrac{15a+1}{15^2 a + 1}$. Since

    $$\begin{aligned}
    A - B &= \frac{a+1}{15a+1} - \frac{15a+1}{15^2 a + 1} = \frac{(a+1)(225a+1) - (15a+1)^2}{(15a+1)(225a+1)} \\
    &= \frac{196a}{(15a+1)(225a+1)} > 0,
    \end{aligned}$$

    so $A > B$, i.e., $\dfrac{15^{2010}+1}{15^{2010}+1} > \dfrac{15^{2011}+1}{15^{2012}+1}$.

3. Since

    $$\log_2[\log_3(\log_4 a)] = 0 \Rightarrow \log_3(\log_4 a) = 1 \Rightarrow \log_4 a = 3 \Rightarrow a = 64,$$
    $$\log_3[\log_4(\log_2 b)] = 0 \Rightarrow \log_4(\log_2 b) = 1 \Rightarrow \log_2 b = 4 \Rightarrow b = 16,$$
    $$\log_4[\log_2(\log_3 c)] = 0 \Rightarrow \log_2(\log_3 c) = 1 \Rightarrow \log_3 c = 2 \Rightarrow c = 9,$$

    so $a + b + c = 89$.

4. Since $x, y, z, w \neq 0$ and $a = 90^{\frac{w}{x}}, b^2 = 90^{\frac{w}{y}}, c = 90^{\frac{w}{z}}$, it follows that

    $$ab^2 c = 90^{\frac{w}{x} + \frac{w}{y} + \frac{w}{z}} = 90 = 2 \cdot 3^2 \cdot 5.$$

    since $a = 1$ implies $w = 0$, so $a > 1$, hence $a = 2, b = 3, c = 5$, the conclusion is proven.

5. Since $7^{\log_{11} 13} = 13^{\log_{11} 7}$,

    $$13^{\log_{11}(x^2 - 10x + 23)} = 7^{\log_{11} 13} \Leftrightarrow 13^{\log_{11}(x^2 - 10x + 23)} = 13^{\log_{11} 7}$$
    $$\Leftrightarrow \log_{11}(x^2 - 10x + 23) = \log_{11} 7 \Leftrightarrow x^2 - 10x + 23 = 7$$
    $$\Leftrightarrow x^2 - 10x + 16 = 0 \Leftrightarrow (x - 2)(x - 8) = 0 \Leftrightarrow x = 2 \text{ or } 8.$$

6.

$$\sqrt{4 + \sqrt{15}} - \sqrt{4 - \sqrt{15}} = \frac{1}{\sqrt{2}} \left(\sqrt{8 + 2\sqrt{15}} - \sqrt{8 - 2\sqrt{15}} \right)$$

$$= \frac{1}{\sqrt{2}} [(\sqrt{5} + \sqrt{3}) - (\sqrt{5} - \sqrt{3})] = \frac{2\sqrt{3}}{\sqrt{2}} = \sqrt{6}.$$

Therefore $\log_2 (\sqrt{4 + \sqrt{15}} - \sqrt{4 - \sqrt{15}}) = \log_2 \sqrt{6} = \frac{1}{2}(1 + \log_2 3).$

7. $2 \lg(x - 2y) = \lg x + \lg y \Leftrightarrow \lg(x - 2y)^2 = \lg(xy) \Leftrightarrow (x - 2y)^2 = xy,$ so

$$x^2 - 5xy + 4y^2 = 0,$$

$$\left(\frac{x}{y} \right)^2 - 5 \left(\frac{x}{y} \right) + 4 = 0,$$

$$\therefore \frac{x}{y} = 1 \ \text{ or } \ 4.$$

However, $x - 2y < 0$ if $x : y = 1$, therefore $x : y = 4$ only, the answer is (A).

8. Let $\log_{b^2} x = u, \log_{x^2} b = v$, then $u + v = 1$ and $x = b^{2u}, b = x^{2v}$, so that

$$x = b^{2u} = (x^{2v})^{2u} = x^{4uv},$$

$$\therefore 4uv = 1, \quad \text{or} \quad uv = \frac{1}{4}.$$

By the inverse Veite's Theorem, (u, v) are the real roots of the equation $w^2 - w + \frac{1}{4} = 0$, therefore $u = v = 1/2$, hence $x = b$, the answer is (D).

9. Since $\dfrac{1}{\log_a b} + \dfrac{1}{\log_b a} = \log_b a + \log_a b$ and $\log_a b \cdot \log_b a = 1$,

$$\left(\frac{1}{\log_{ab} b} - \frac{1}{\log_{ab} a} \right)^2 = (\log_b ab - \log_a ab)^2 = (\log_b a - \log_a b)^2$$

$$= (\log_b a)^2 + (\log_a b)^2 - 2 = (\log_b a + \log_a b)^2 - 4$$

$$= 1229 - 4 = 1225.$$

Besides, $a > b > 1 \Rightarrow \log_{ab} b < \log_{ab} a \Rightarrow \dfrac{1}{\log_{ab} b} - \dfrac{1}{\log_{ab} a} > 0$, so

$$\frac{1}{\log_{ab} b} - \frac{1}{\log_{ab} a} = \sqrt{1225} = 35.$$

10. The given equation gives $(3^x - 9)^3 + (9^x - 81)^3 + [-(9^x + 3^x - 90)]^3 = 0$, so
$$3(3^x - 9)(9^x - 81)(9^x + 3^x - 90) = 0.$$

When $3^x - 9 = 0$, then $x = 2$; when $9^x - 81 = 0$, then $x = 2$ also; when $9^x + 3^x - 90 = 0$, then
$$(3^x - 9)(3^x + 10) = 0$$

which also implies that $x = 2$ also. Thus, the sum of real roots is 2.

11. Taking the operation $\log_5$ to both sides of the equation, then
$(\log_5 x)(\log_5 15) + (\log_5 45x)(\log_5 x) = 0$
$\Rightarrow \log_5 x(\log_5 15 + \log_5 45 + \log_5 x) = 0$
$\Rightarrow \log_5 x(\log_5 675 + \log_5 x) = 0 \Rightarrow \log_5 x = 0$ or $\log_5 675 + \log_5 x = 0$
$\Rightarrow x = 1$ or $x = \dfrac{1}{675}.$

Testing Questions   (4-B)

1. By setting $2^x = a$ and $3^x = b$, the equation becomes
$$\frac{a^3 + b^3}{a^2 b + ab^2} = \frac{7}{6} \Leftrightarrow \frac{a^2 - ab + b^2}{ab} = \frac{7}{6}$$
$$\Leftrightarrow 6a^2 - 13ab + 6b^2 = 0 \Leftrightarrow (2a - 3b)(3a - 2b) = 0.$$

Therefore $2^{x+1} = 3^{x+1}$ or $2^{x-1} = 3^{x-1}$, which implies that $x_1 = -1$ and $x_2 = 1$.

It is easy to check that both $x = -1$ and $x = 1$ satisfy the given equation.

2.
$$\begin{aligned}
f(x) &= a^x + b^x - c^x - \left(\frac{ab}{c}\right)^x + d^x \left(\frac{ab}{cd}\right)^x - d^x \\
&= c^x\left[\left(\frac{a}{c}\right)^x - 1\right] + b^x\left[1 - \left(\frac{a}{c}\right)^x\right] + d^x\left[\left(\frac{ab}{cd}\right)^x - 1\right] \\
&= b^x\left[\left(\frac{a}{c}\right)^x - 1\right]\left[\left(\frac{c}{b}\right)^x - 1\right] + d^x\left[\left(\frac{ab}{cd}\right)^x - 1\right].
\end{aligned}$$

Since $b > 1, \dfrac{a}{c} > 1, \dfrac{c}{b} > 1, d > 1, \dfrac{ab}{cd} > 1$, all the factors in the last expression are positive and strictly increasing, so f is strictly increasing.

3. First of all,

$$\sum_{k=2}^{n} [\log_{\frac{3}{2}} (k^3 + 1) - \log_{\frac{3}{2}} (k^3 - 1)]$$

$$= \sum_{k=2}^{n} \log_{\frac{3}{2}} \frac{k^3 + 1}{k^3 - 1} = \log_{\frac{3}{2}} \prod_{k=2}^{n} \frac{k^3 + 1}{k^3 - 1} = \log_{\frac{3}{2}} \prod_{k=2}^{n} \frac{(k + 1)(k^2 - k + 1)}{(k - 1)(k^2 + k + 1)}$$

$$= \log_{\frac{3}{2}} \left(\prod_{k=2}^{n} \frac{k + 1}{k - 1} \cdot \prod_{k=2}^{n} \frac{k^2 - k + 1}{k^2 + k + 1} \right).$$

From

$$\prod_{k=2}^{n} \frac{k + 1}{k - 1} = \frac{3}{1} \cdot \frac{4}{2} \cdot \frac{5}{3} \cdots \frac{n - 1}{n - 3} \cdot \frac{n}{n - 2} \cdot \frac{n + 1}{n - 1} = \frac{n(n + 1)}{2}$$

and

$$\prod_{k=2}^{n} \frac{k^2 - k + 1}{k^2 + k + 1} = \frac{3}{7} \cdot \frac{7}{13} \cdot \frac{13}{17} \cdots \frac{n^2 - n + 1}{n^2 + n + 1} = \frac{3}{n^2 + n + 1}$$

since $(k - 1)^2 + (k - 1) + 1 = k^2 - k + 1$, it is obtained that

$$\sum_{k=2}^{n} [\log_{\frac{3}{2}} (k^3 + 1) - \log_{\frac{3}{2}} (k^3 - 1)] = \log_{\frac{3}{2}} \left[\frac{n(n + 1)}{2} \cdot \frac{3}{n^2 + n + 1} \right]$$

$$= \log_{\frac{3}{2}} \left(\frac{3}{2} \cdot \frac{n^2 + n}{n^2 + n + 1} \right) < \log_{\frac{3}{2}} \frac{3}{2} = 1.$$

4. For $x < 0$, the function $f(x) = 2^x + 3^x + 6^x - x^2$ is increasing, so the equation $f(x) = 0$ has the unique solution $x = -1$.

Assume that there is a solution $s \geq 0$. Then $s^2 = 2^s + 3^s + 6^s \geq 3$, so $s \geq \sqrt{3}$, and hence $\lfloor s \rfloor \geq 1$.

But $s \geq \lfloor s \rfloor \geq 1$ yields

$$2^s \geq 2^{\lfloor s \rfloor} = (1 + 1)^{\lfloor s \rfloor} \geq 1 + \lfloor s \rfloor > s,$$

which in turn implies that

$$6^s > 4^s = (2^s)^2 > s^2.$$

So $2^s + 3^s + 6^s > s^2$, a contradiction. Thus, $x = -1$ is the only solution.

5. (4.4) yields

$$y = \frac{1}{x^2} \tag{15.29}$$

and (4.3) yields

$$\left(\frac{x}{y}\right)^x = \left(\frac{1}{xy}\right)^y. \tag{15.30}$$

Then substituting (15.29) into (15.30) gives $x^{3x} = x^y$ or $x^{3x-y} = 1$. When $x = \pm 1$, (15.29) gives $y = 1$. By checking, $(-1, 1)$ and $(1, 1)$ are roots of the original system.

When $x \neq -1$ and $x \neq 1$, then $y = 3x$, so $x^3 = \frac{1}{3}$ or $x = \frac{1}{\sqrt[3]{3}}$ by (15.29), and hence $y = \sqrt[3]{9}$.

Thus, the solutions of the given system are $(-1, 1)$, $(1, 1)$ and $\left(\frac{1}{\sqrt[3]{3}}, \sqrt[3]{9}\right)$.

6. It's clear that $0 \leq \lfloor \log_2 k \rfloor \leq 8$ for $1 \leq k \leq 256$. For each $i \in \{0, 1, 2, \ldots, 7\}$

$$\lfloor \log_2 k \rfloor = i \Leftrightarrow 2^i \leq k < 2^{i+1} \Leftrightarrow k = 2^i, 2^i + 1, \ldots, 2^{i+1} - 1,$$

so there are 2^i of k such that $\lfloor \log_2 k \rfloor = i$. Therefore

$$\lfloor \log_2 1 \rfloor + \lfloor \log_2 2 \rfloor + \lfloor \log_2 3 \rfloor + \cdots + \lfloor \log_2 256 \rfloor$$
$$= 0 \cdot 1 + 1 \cdot 2^1 + 2 \cdot 2^2 + 3 \cdot 2^3 + 4 \cdot 2^4 + 5 \cdot 2^5 + 6 \cdot 2^6 + 7 \cdot 2^7 + 8$$
$$= 1546.$$

7. $a > 0, a^2 - 1 > 0, a^x - 1 \neq 1$ for any $x \in (0, 1]$ yields $a > 1, a \neq 2$ and $x \neq \log_a 2$ respectively. Then the given inequality holds for any $x \in (0, 1]$ means that

$$\log_a 2 > 1,$$

so $a < 2$. On the other hand,

$$\log_a(a^x + 1) + \frac{1}{\log_{a^{x-1}} a} \leq x - 1 + \log_a(a^2 - 1)$$
$$\Leftrightarrow \log_a(a^x + 1)(a^x - 1) \leq \log_a a^{x-1}(a^2 - 1)$$
$$\Leftrightarrow (a^x - 1)(a^x + 1) \leq a^{x+1} - a^{x-1}$$
$$\Leftrightarrow a^{2x} - a^{x+1} + a^{x-1} - 1 \leq 0 \Leftrightarrow (a^{x+1} + 1)(a^{x-1} - 1) \leq 0.$$

Since $a^{x+1} + 1 > 0$ for $a > 1, 0 < x \leq 1$ and for any $0 < x \leq 1$

$$a^{x-1} - 1 \leq 0 \Leftrightarrow a > 1,$$

Thus, $1 < a < 2$.

Solutions to Testing Questions 5

Testing Question (5-A)

1. For $a > 1$, the function $y = a^x$ is an increasing function. For $0° < \theta < 45°$, $\cot \theta > 1 > \tan \theta > 0$. Thus $t_3 < t_4$.

   For $a < 1$, the function $y = a^x$ is a decreasing function. Thus $t_1 > t_2$.

   Again, by $\cot \theta > 1 > \tan \theta > 0$, we have $t_1 < 1 < t_3$. Hence $t_4 > t_3 > t_1 > t_2$.

2. Let $\alpha_1 = \dfrac{\pi}{2} - \sin x$, $\alpha_2 = \cos x$.

   (i) When $0 \leq x \leq \dfrac{\pi}{2}$, then $0 < \alpha_1 \leq \frac{\pi}{2}, 0 \leq \alpha_2 \leq 1$ and

   $$\alpha_1 - \alpha_2 = \frac{\pi}{2} - (\sin x + \cos x) \geq \frac{\pi}{2} - \sqrt{2} > 0,$$

   so $\cos(\sin x) = \sin \alpha_1 > \sin \alpha_2 = \sin(\cos x)$.

   (ii) When $\dfrac{\pi}{2} < x < \pi$, then $-1 < \cos x < 0$, so $\sin(\cos x) < 0$, whereas

   $$0 < \sin x < 1 \Rightarrow \cos(\sin x) > 0,$$

   so $\cos(\sin x) > \sin(\cos x)$ for $\dfrac{\pi}{2} < x < \pi$. When $x = \dfrac{\pi}{2}$, then $\cos(\sin \frac{\pi}{2}) = \cos 1 > 0 = \sin(\cos \frac{\pi}{2})$. When $x = \pi$, then $\cos(\sin \pi) = \cos 0 = 1 > -\sin 1 = \sin(\cos \pi)$.

   Thus, the conclusion is proven.

3. From assumption it is obtained that $9 = a \tan 7 - b \sin 7 + 2$, so

   $$a \tan 7 - b \sin 7 = 7.$$

   Then

   $$f(-7) = a \tan(-7) - b \sin(-7) + 2 = -(a \tan 7 - b \sin 7) + 2 = -7 + 2 = -5.$$

4. Since $0 < \sin x < \cos x < 1$ for $0 < x < \dfrac{\pi}{4}$,

   $$\cos x^{\sin x} > \sin x^{\sin x} > \sin x^{\cos x},$$

   therefore $\cos x^{\sin x^{\sin x}} < \cos x^{\sin x^{\cos x}}$, hence $a < c$.

Similarly, $\cos x^{\sin x} > \sin x^{\sin x}$ gives

$$\sin x^{\cos x^{\sin x}} < \sin x^{\sin x^{\sin x}} < \cos x^{\sin x^{\sin x}},$$

therefore $b < d < a$. Thus, $b < d < a < c$.

5. $f(x + 2\pi) = \cos(\sin(x + 2\pi)) = \cos(\sin x) = f(x)$ shows that 2π is a period of f.

Suppose that $T \in (0, 2\pi)$ is another period of f. Then

$$\cos(\sin(x + T)) = f(x + T) = f(x) = \cos(\sin x), \qquad \forall x \in \mathbb{R}.$$

Letting $x = 0$ gives $\cos(\sin T) = \cos(\sin 0) = 1$, so $\sin T = 2k\pi$. Since $0 < T < 2\pi$, so $k = 0$ and $T = \pi$. Then

$$\cos(\sin(x + \pi)) = \cos(-(\sin x)) = \cos(\sin x).$$

Thus, the minimum period of f is π.

6. Suppose that $b = ar, c = ar^2$, where $r > 0$ is the common ratio of the G.P.. Then

$$\frac{\sin A \cot C + \cos A}{\sin B \cot C + \cos B} = \frac{\sin A \cos C + \cos A \sin C}{\sin B \cos C + \cos B \sin C}$$

$$= \frac{\sin(A + C)}{\sin(B + C)} = \frac{\sin B}{\sin A} = \frac{b}{a} = r.$$

Since a, b, c form a G.P, so their maximum value is a or c. By the triangle inequality, $a + b > c$ and $b + c > a$. Then

$$a + ar > ar^2 \Rightarrow r^2 - r - 1 < 0, \quad \text{and} \quad ar + ar^2 > a \Rightarrow r^2 + r - 1 > 0.$$

The solution set the first inequality is $0 < r < \dfrac{1 + \sqrt{5}}{2}$, and the solution set of the second inequality is $r > \dfrac{\sqrt{5} - 1}{2}$. Thus, $\dfrac{\sqrt{5} - 1}{2} < r < \dfrac{\sqrt{5} + 1}{2}$, and the range of the given expression is

$$\left(\frac{\sqrt{5} - 1}{2}, \frac{\sqrt{5} + 1}{2} \right).$$

7. By simplification,

$$\begin{aligned}
f(x) &= \sin^4 x - \sin x \cos x + \cos^4 x \\
&= (\sin^2 x + \cos^2 x)^2 - \frac{1}{2}\sin^2 2x - \frac{1}{2}\sin 2x \\
&= 1 - \frac{1}{2}\sin 2x - \frac{1}{2}\sin^2 2x.
\end{aligned}$$

Let $t = \sin 2x$, then $-1 \le t \le 1$, and

$$f(x) = g(t) = 1 - \frac{1}{2}t - \frac{1}{2}t^2 = \frac{9}{8} - \frac{1}{2}\left(t + \frac{1}{2}\right)^2.$$

Therefore $\min_{x \in \mathbb{R}}\{f(x)\} = \min_{-1 \le t \le 1}\{g(t)\} = 0$, and $\max_{x \in \mathbb{R}}\{f(x)\} = \max_{-1 \le t \le 1}\{g(t)\} = \frac{9}{8}$. Thus, the range of f is $\left[0, \frac{9}{8}\right]$.

8. Let $t = \sin\theta + \cos\theta$, then $\cos\left(\theta - \frac{\pi}{4}\right) = \frac{\sqrt{2}}{2}t$, $\sin 2\theta = t^2 - 1$. Since $0 < t \le \sqrt{2}$ and $\sin 2\theta \ge 0$, so $1 \le t \le \sqrt{2}$. The given inequality then becomes

$$(2a + 3)t + \frac{6}{t} - 2(t^2 - 1) < 3a + 6$$

or, equivalently, $2t^2 - 2at - 3t - \frac{6}{t} + 3a + 4 > 0$. Thus,

$$2t\left(t + \frac{2}{t} - a\right) - 3\left(t + \frac{2}{t} - a\right) > 0,$$

$$(2t - 3)\left(t + \frac{2}{t} - a\right) > 0, \quad t \in [1, \sqrt{2}],$$

$$\therefore t + \frac{2}{t} < a \quad \text{for any } t \in [1, \sqrt{2}].$$

Since $f(t) = t + \frac{2}{t}$ on $[1, \sqrt{2}]$ is decreasing, so it suffices to let $a > f(1) = 3$.

Thus, the range of a is $(3, +\infty)$.

9. Suppose that $T > 0$ is a period, then $f(0) = 0 = f(nT)$ for $n = 0, 1, 2, 3, \ldots$. However, for any $n > \frac{3}{T}$,

$$|f(nT)| = |-nT + \sin(nT)| \ge |-nT| - |\sin(nT)| > 3 - 1 = 2,$$

a contradiction.

10 Since $2008° = 5 \times 360° + 180° + 28°$, so

$$a = \sin(-\sin 28°) = -\sin(\sin 28°) < 0;$$
$$b = \sin(-\cos 28°) = -\sin(\cos 28°) < 0;$$
$$c = \cos(-\sin 28°) = \cos(\sin 28°) > 0;$$
$$d = \cos(-\cos 28°) = \cos(\cos 28°) > 0.$$

Since $\sin 28° < \cos 28°$, so $b < a < d < c$, the answer is (B).

Testing Questions   (5-B)

1. Let T_1 be a period which is not a multiple of T_0, then $T_1 = mT_0 + T'$, where $m \in \mathbb{Z}, 0 < T' < T_0$. Then

$$f(x + T') = f(x + mT_0 + T') = f(x + T_1) = f(x),$$

i.e., T' is also a period of f, which contradicts that T_0 is the minimal positive period of f. Thus, the conclusion is proven.

2. The answers are $\alpha = \dfrac{\pi}{8} + \dfrac{k\pi}{2}$ for all integers k.

Because $S = T$, the sums of the elements in S and T are equal to each other; that is,

$$\sin \alpha + \sin 2\alpha + \sin 3\alpha = \cos \alpha + \cos 2\alpha + \cos 3\alpha.$$

Applying the sum-to-product formulas to the first and the third summands on each side of the last equation gives

$$2 \sin 2\alpha \cos \alpha + \sin 2\alpha = 2 \cos 2\alpha \cos \alpha + \cos 2\alpha,$$

or

$$\sin 2\alpha (2 \cos \alpha + 1) = \cos 2\alpha (2 \cos \alpha + 1).$$

If $2 \cos \alpha + 1 = 0$, then $\cos \alpha = -\frac{1}{2}$, and so $\alpha = \pm\frac{2\pi}{3} + 2k\pi$ for all integers k. It is then not difficult to check that $S \neq T$ since $\sin 3\alpha = 0$ but $0 \notin \{\cos \alpha, \cos 2\alpha, \cos 3\alpha\}$.

Thus, $2 \cos \alpha + 1 \neq 0 \Rightarrow \sin 2\alpha = \cos 2\alpha \Rightarrow \tan 2\alpha = 1 \Rightarrow \alpha = \frac{\pi}{8} + \frac{k\pi}{2}$ for all integers k. It is not difficult to check that in this case we have $S = T$: Since $3\alpha + \alpha = \frac{\pi}{2} + 2k\pi$, so $\cos 3\alpha = \sin \alpha, \cos \alpha = \sin 3\alpha$.

3. For the sake of contradiction, suppose that $T > 0$ is a period of f. Then

$$f(T) = f(0) = 1 = f(\sqrt[4]{2k\pi}) = f(\sqrt[4]{2k\pi} + T), \quad k \in \mathbb{Z}.$$

Therefore there exist an integer $m \neq 0$ such that $T^4 = 2m\pi$ or $T = \sqrt[4]{2m\pi}$. Similarly, for $k = 2m$, there exists $n \in \mathbb{Z}$ such that

$$(\sqrt[4]{4m\pi} + T)^4 = 2n\pi.$$

Thus,

$$(\sqrt[4]{4m\pi} + \sqrt[4]{2m\pi})^4 = 2n\pi \Rightarrow 2m\pi(\sqrt[4]{2} + 1)^4 = 2n\pi$$
$$\Rightarrow m(2 + 4\sqrt[4]{8} + 6\sqrt{2} + 4\sqrt[4]{2} + 1) = n \Rightarrow 2\sqrt[4]{8} + 3\sqrt{2} + 2\sqrt[4]{2} \in \mathbb{Q}.$$

Write $a = \sqrt[4]{2}$, then $A = 2a^3 + 3a^2 + 2a \in \mathbb{Q}$. However, from $a^4 = 2$

$A^2 \in \mathbb{Q} \Rightarrow a^2(4a^4 + 9a^2 + 4 + 12a^3 + 12a + 8a^2) \in \mathbb{Q}$
$\Rightarrow 12a^3 + 12a^2 + 24a \in \mathbb{Q} \Rightarrow 6(2a^3 + 3a^2 + 2a) - 6(a^2 - 2a) \in \mathbb{Q}$
$\Rightarrow a^2 - 2a \in \mathbb{Q}.$

Let $B = a^2 - 2a \in \mathbb{Q}$. Then

$$B^2 \in \mathbb{Q} \Rightarrow 4a^2 - 4a^3 \in \mathbb{Q} \Rightarrow 2a^3 - 2a^2 \in \mathbb{Q} \Rightarrow -5a^2 - 2a \in \mathbb{Q}.$$

Combining it with $B = a^2 - 2a \in \mathbb{Q}$, it follows that $a^2 = \sqrt{2} \in \mathbb{Q}$, a contradiction.

4.   First of all, $2008 = 2^{10} + 2^9 + 2^8 + 2^7 + 2^6 + 2^4 + 2^3$, so

$$\sum_{i=1}^{n} \alpha_i = 10 + 9 + 8 + 7 + 6 + 4 + 3 = 47.$$

Since $3.14 < \pi < 15$, $15\pi > 3.14 \times 15 > 47$, and $(14\frac{5}{6})\pi = \frac{89}{6}\pi < 46.72 < 47$,

$$14\pi + \frac{5}{6}\pi < 47 < 15\pi,$$

therefore $\sin \sum\limits_{i=1}^{n} \alpha_i = \sin 47 > 0$. Since $y = \cos x$ is decreasing but $\tan x$ is increasing on $[14\pi + \frac{5}{6}\pi, 15\pi]$, so

$$\cos 47 < \cos\left(14\pi + \frac{5}{6}\pi\right) = -\frac{\sqrt{3}}{2},$$
$$0 > \tan 47 > \tan\left(14\pi + \frac{5}{6}\pi\right) = -\frac{\sqrt{3}}{3}.$$

Thus, $\sin \sum\limits_{i=1}^{n} \alpha_i > \tan \sum\limits_{i=1}^{n} \alpha_i > \cos \sum\limits_{i=1}^{n} \alpha_i.$

5.   It is clear that the function $f(x) = 0$ identically on $\mathbb{R}$ satisfies the requirement in question.

Now consider that f is not the zero polynomial. If f satisfies the given requirement, then

$f(\sin \pi(x + 2)) = f(\sin(\pi x + 2\pi)) = f(\sin \pi x) = f(x)\cos \pi x,$
$f(x + 2)\cos \pi(x + 2) = f(x + 2)\cos(\pi x + 2\pi) = f(x + 2)\cos \pi x,$

therefore $f(x+2)\cos \pi x = f(x)\cos \pi x$. When $x \neq k + \frac{1}{2}$ where $k \in \mathbb{Z}$, then $f(x+2) = f(x)$. Thus, it suffices to determine f on $[-1, 1]$.

Let $x = \frac{1}{2}$, then given equality gives $f(1) = f(\frac{1}{2})\cos \frac{\pi}{2} = 0$.

Let $x = 1$, then the given equality gives $f(0) = f(1)\cos \pi = 0$.

On $[-1, 1]$, $|f|$ is continuous, so its maximum value M is reachable, i.e., there exists $x_0 \in [-1, 1]$ such that $M = |f(x_0)|$. Let $y_0 = \frac{1}{\pi}\sin^{-1} x_0$,

i.e., $-\frac{1}{2} \leq y_0 \leq \frac{1}{2}$ and $\sin \pi y_0 = x_0$, then

$$M = |f(x_0)| = |f(\sin \pi y_0)| = |f(y_0)| \cdot |\cos \pi y_0| \leq |f(y_0)|,$$

hence $|f(y_0)| = M$ and $|\cos \pi y_0| = 1$, i.e., y_0 is an integer. From $|y_0| \leq \frac{1}{2}$ it follows that $y_0 = 0$. Thus, $M = |f(0)| = 0$, i.e. $f(0) = 0$ identically on $[-1, 1]$.

Thus, $f(x) = 0$ identically on $\mathbb{R}$ is the unique solution.

Solutions to Testing Questions   (6)

Testing Questions   (6-A)

1. Let $AB = AC = 1, \angle C = 2\alpha$, as shown in the given graph below, then

$\angle ABD = \alpha, \angle BDA = 3\alpha, \angle A = 180° - 4\alpha$.

Applying the sine rule to $\triangle ABC$, then

$$\frac{BC}{\sin 4\alpha} = \frac{1}{\sin 2\alpha} \Rightarrow BC = 2\cos 2\alpha.$$

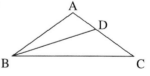

Applying the sine rule to $\triangle ABD$ gives $\dfrac{1}{\sin 3\alpha} = \dfrac{BD}{\sin 4\alpha} = \dfrac{AD}{\sin \alpha}$, so $BD = \dfrac{\sin 4\alpha}{\sin 3\alpha}, AD = \dfrac{\sin \alpha}{\sin 3\alpha}$. Then

$$BC = BD + AD \Rightarrow 2\cos 2\alpha = \frac{\sin 4\alpha}{\sin 3\alpha} + \frac{\sin \alpha}{\sin 3\alpha}$$
$$\Rightarrow 2\sin 3\alpha \cos 2\alpha = \sin 4\alpha + \sin \alpha \Rightarrow \sin 5\alpha + \sin \alpha = \sin 4\alpha + \sin \alpha$$
$$\Rightarrow \sin 5\alpha = \sin 4\alpha \Rightarrow 5\alpha = 180° - 4\alpha \Rightarrow \alpha = 20°.$$

Thus, $\angle A = 180° - 80° = 100°$.

2. Apply the cosine rule to $\triangle BDC$ and $\triangle ABC$ respectively, it follows that

$$BD^2 = BC^2 + CD^2 - 2BC \cdot CD \cos C, \qquad (15.31)$$
$$AB^2 = BC^2 + AC^2 - 2BC \cdot AC \cos C. \qquad (15.32)$$

Let $CD = x$. From (15.31),

$$CD^2 - (14 \cos C) \cdot CD + 33 = 0,$$

and similarly from (15.32),

$$AC^2 - (14 \cos C) \cdot AC + 33 = 0.$$

Therefore AC, CD are the roots of the equation $x^2 - (14 \cos C)x + 33 = 0$, hence $AC \cdot CD = 33$ i.e. $CD = \dfrac{33}{10}$. Thus, $AD = 10 - \dfrac{33}{10} = \dfrac{67}{10}$ and

$$AD : DC = 67 : 33.$$

3. Extend the ray CB to B' and the ray BC to C', such that $BA = BB'$ and $CA = CC'$, as shown in the following graph,

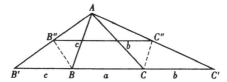

then triangles ABB' and ACC' are both isosceles, $\angle B' = \frac{1}{2}\angle B$, $\angle C' = \frac{1}{2}\angle C$.

Let B'', C'' be the midpoints of AB', AC' respectively. In $\triangle AB''C''$,

$$AB'' = c \cos \frac{B}{2}, \quad AC'' = b \cos \frac{C}{2}, \quad B''C'' = p,$$

and

$$\angle B''AC'' = \angle A + \frac{\angle B}{2} + \frac{\angle C}{2} = \pi - \frac{\angle B + \angle C}{2}.$$

Then applying the cosine rule to $\triangle AB''C''$ gives

$$B''C''^2 = AC''^2 + AB''^2 - 2AB'' \cdot AC'' \cdot \cos \angle B''AC'',$$

i.e,

$$p^2 = b^2 \cos^2 \frac{C}{2} + c^2 \cos^2 \frac{B}{2} + 2bc \cos \frac{B}{2} \cos \frac{C}{2} \cos \frac{B+C}{2},$$

as desired.

4. Since $2 = [ABC] = \frac{1}{2}bc \sin A = \frac{\sqrt{2}}{2}c$, it follows that $c = 2\sqrt{2}$. On the
    other hand, the cosine rule gives

    $$a^2 = b^2 + c^2 - 2bc \cos A = 4 + 8 - 8\sqrt{2} \cdot \frac{\sqrt{2}}{2} = 4,$$

    therefore $a = 2$. By using the sine rule,

    $$\frac{a + b + c}{\sin A + \sin B + \sin C} = \frac{a}{\sin A} = 2 : \frac{\sqrt{2}}{2} = 2\sqrt{2}.$$

5. It suffices to consider the following two cases:

    (i) When M is between A and C, then

    $$\angle AMB = \angle ADB = 2\angle C \Rightarrow \angle MBC = \angle C.$$

    Let r_1, r_2 be the radii of the circumcircles of
    $\triangle MNC$ and $\triangle ABD$ respectively, then

    $$2r_1 \sin \angle C = MN = 2r_2 \sin \angle MBC,$$
    $$\therefore r_1 = r_2.$$

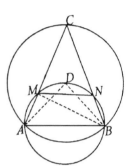

    (ii) When M is on the extension of CA, then

    $$\angle CMB = \pi - \angle ADB = \pi - 2\angle C,$$
    $$\therefore \angle MBC = \angle C,$$

    hence, $2r_1 \sin C = MN = 2r_2 \sin \angle MBC$
    gives also the desired conclusion

    $$r_1 = r_2.$$

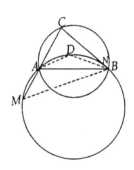

6. Let $CA = b$, $AB = c$. By applying the cosine rule to $\triangle BCK$, then

    $$\cos C = \frac{4 + 1 - \frac{9}{2}}{2 \cdot 2 \cdot 1} = \frac{1}{8}.$$

    By applying the cosine rule to $\triangle ABC$,

    $$\frac{1}{8} = \cos C = \frac{4 + b^2 - c^2}{4b}.$$

    Therefore $8 + 2b^2 - 2c^2 = b$. The angle
    bisector theorem gives $b - 1 = \frac{c}{2}$, i.e., $c = 2(b - 1)$

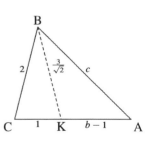

By solving the system of two equations, it's easy to find that $b = \dfrac{5}{2}, c = 3$. By Heron's Formula,

$$[ABC] = \sqrt{\frac{15}{4} \cdot \frac{7}{4} \cdot \frac{5}{4} \cdot \frac{3}{4}} = \frac{15\sqrt{7}}{16}.$$

7. Let $R = \dfrac{\sqrt{3}}{6}$ be the circumradius. From the extended sine rule,

$$\sin A = 2a \sin B \Rightarrow \frac{a}{2R} = \frac{2ab}{2R} \Rightarrow b = \frac{1}{2}.$$

Therefore

$$\sin B = \frac{b}{2R} = \frac{\frac{1}{2}}{2 \cdot \frac{\sqrt{3}}{6}} = \frac{\sqrt{3}}{2}.$$

Since $\angle A, \angle C, \angle B < \dfrac{\pi}{2}$, $\angle B = \dfrac{\pi}{3}$, and

$$\angle A = \angle(A + C) - \angle C = \frac{2\pi}{3} - \angle C > \frac{2\pi}{3} - \frac{\pi}{2} = \frac{\pi}{6},$$

so $\angle A \in (\frac{\pi}{6}, \frac{\pi}{2})$. Thus, the perimeter l of the $\triangle ABC$ is given by

$$
\begin{aligned}
l &= a + b + c = 2R \sin A + 2R \sin C + b = \frac{1}{2} + \frac{\sqrt{3}}{3}(\sin A + \sin C) \\
&= \frac{1}{2} + \frac{\sqrt{3}}{3}\left(\sin A + \sin\left(\frac{2\pi}{3} - A\right)\right) \\
&= \frac{1}{2} + \frac{\sqrt{3}}{3}\left(\sin A + \frac{\sqrt{3}}{2}\cos A + \frac{1}{2}\sin A\right) \\
&= \frac{1}{2} + \left(\frac{\sqrt{3}}{2}\sin A + \frac{1}{2}\cos A\right) \\
10pt] \quad &= \frac{1}{2} + \sin\left(A + \frac{\pi}{6}\right).
\end{aligned}
$$

Since the range of $A + \frac{\pi}{6}$ is $\left(\dfrac{\pi}{3}, \dfrac{2\pi}{3}\right)$, so that

$$\frac{\sqrt{3}}{2} < \sin\left(A + \frac{\pi}{6}\right) \le 1,$$

the range of l is $\left(\dfrac{1+\sqrt{3}}{2}, \dfrac{3}{2}\right]$.

8.   By the extended sine rule,

$$R = \frac{a}{2\sin A} = \frac{abc}{2bc\sin A} = \frac{abc}{4[ABC]},$$

establishing (a). By the same token, we have

$$\begin{aligned} 2R^2 \sin A \sin B \sin C &= \frac{1}{2} \cdot (2R\sin A)(2R\sin B)(\sin C) \\ &= \frac{1}{2}ab\sin C = [ABC], \end{aligned}$$

which is (b).

Note that
$$2[ABC] = bc\sin A = (a+b+c)r.$$

By the extended sine rule also,

$$\begin{aligned} 4R^2 \sin A \sin B \sin C &= bc\sin A = r(a+b+c) \\ &= 2rR(\sin A + \sin B + \sin C), \end{aligned}$$

from which (c) follows.

Testing Questions   (6-B)

1.   From Sine rule and Cosine rule, the first equality can be written in the form $\dfrac{\sin\beta}{\sin\alpha} = 2|\cos\alpha|$, i.e. $\sin\beta = |\sin 2\alpha|$. It is equivalent to

(i) $\beta = 2\alpha$ or $\beta = 180° - 2\alpha$ or $\beta = 2\alpha - 180°$.

Similarly, the rest two equalities are equivalent to

(ii) $\gamma = 2\beta$ or $\gamma = 180° - 2\beta$ or $\gamma = 2\beta - 180°$;

(iii) $\alpha = 2\gamma$ or $\alpha = 180° - 2\gamma$ or $\alpha = 2\gamma - 180°$.

Since $\alpha + \beta + \gamma = 180°$, if $\beta = 180° - 2\alpha$, then $\alpha = \gamma$. In this case, only $\alpha = 180° - 2\gamma$ is possible, so $\alpha = \beta = \gamma = 60°$. We get the same conclusion if $\gamma = 180° - 2\beta$ or $\alpha = 180° - 2\gamma$.

When above three equalities do not hold, we may assume that α is the smallest one in the three angles, then $\beta = 2\alpha - 180°$ is impossible, so $\beta = 2\alpha$.

If $\gamma = 2\beta$, then $\alpha : \beta : \gamma = 1 : 2 : 4$, this is the second possible solution.

If $\gamma = 2\beta - 180°$, then $\gamma = 4\alpha - 180°$, from $\alpha + \beta + \gamma = 180°$ we obtain

$$\alpha = \frac{360°}{7}, \quad \beta = 2 \times \frac{360°}{7}, \quad \gamma = 4 \times \frac{360°}{7} - 180° = \frac{180°}{7} < \alpha,$$

a contradiction. Therefore it is impossible.

Thus, $\alpha = \beta = \gamma = 60°$, or the ratios of these three angles are $1 : 2 : 4$.

2. By the cosine rule,

$$\cos B = \frac{7^2 + 8^2 - 9^2}{2 \cdot 7 \cdot 8} = \frac{2}{7},$$

$$\cos \frac{B}{2} = \left(\frac{1 + \cos B}{2} \right)^{\frac{1}{2}} = \sqrt{\frac{9}{14}}.$$

Let the perpendiculars from D to the lines BC, CA and BA intersect them at E, F and G respectively, then

$$CF + AF = 9, AF - CF = 8 - 7 = 1 \Rightarrow AF = 5, CF = 4,$$

therefore $BG = BE = 12$ and

$$BD^2 = \left(\frac{BG}{\cos \frac{B}{2}} \right)^2 = \frac{12^2 \cdot 14}{9} = 224.$$

3. Let R be the circumradius of the quadrilateral. The sine rule gives

$$AC = 2R \sin(\alpha_4 + \alpha_1) = 2R \sin(\alpha_2 + \alpha_3),$$
$$BD = 2R \sin(\alpha_1 + \alpha_2) = 2R \sin(\alpha_3 + \alpha_4),$$
$$AB = 2R \sin \alpha_3, \quad BC = 2R \sin \alpha_2,$$
$$CD = 2R \sin \alpha_1, \quad DA = 2R \sin \alpha_4.$$

Substututing them into the inequality to be proven, it becomes

$$AC^2 \cdot BD^2 \geq 4AB \cdot BC \cdot CD \cdot DA. \qquad (*)$$

Applying the Ptolemy's Theorem to $ABCD$ gives

$$AC \cdot BD = AB \cdot CD + BC \cdot AD,$$

so that the AM-GM inequality yields

$$AC^2 \cdot BD^2 = (AB \cdot CD + BC \cdot AD)^2 \geq 4AB \cdot BC \cdot CD \cdot DA,$$

as desired.

4. Let $BC = a, CA = b, AB = c$. We may assume that $\alpha > \beta$, as shown in the given graph.

    Suppose that the rays DE and BA intersect at F, and the included angle is φ. Then the angle bisector theorem gives

    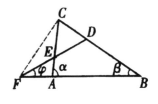

    $$\frac{BD}{DC} = \frac{c}{b} \text{ and } \frac{CE}{EA} = \frac{a}{c} \Rightarrow BD = \frac{ac}{b+c},$$

    $$DC = \frac{ab}{b+c}, CE = \frac{ab}{a+c}, AE = \frac{bc}{a+c}.$$

    Applying the Menelaus' Theorem to $\triangle ABC$ and the transversal FD yields

    $$\frac{AF}{FB} \cdot \frac{BD}{DC} \cdot \frac{CE}{EA} = 1,$$

    so $\dfrac{AF}{AF+c} = \dfrac{b}{a} \Rightarrow AF = \dfrac{bc}{a-b}, FB = \dfrac{ac}{a-b}.$

    Applying the sine rule to $\triangle AEF$ and $\triangle FDB$ gives

    $$\frac{\sin(\alpha - \varphi)}{\sin\varphi} = \frac{\sin \angle AEF}{\sin \angle AFE} = \frac{AF}{AE} = \frac{bc/(a-b)}{bc/(a+c)} = \frac{a+c}{a-b},$$

    $$\frac{\sin(\beta + \varphi)}{\sin\varphi} = \frac{\sin \angle FDB}{\sin \angle BFD} = \frac{FB}{BD} = \frac{ac/(a-b)}{ac/(b+c)} = \frac{b+c}{a-b},$$

    therefore $\dfrac{\sin(\alpha - \varphi) - \sin(\beta + \varphi)}{\sin\varphi} = \dfrac{a+c}{a-b} - \dfrac{b+c}{a-b} = 1$, so that

    $$\sin\varphi = \sin(\alpha - \varphi) - \sin(\beta + \varphi) = 2 \sin \frac{\alpha - \beta - 2\varphi}{2} \cdot \cos \frac{\alpha + \beta}{2}$$

    $$< 2 \sin \frac{\alpha - \beta - 2\varphi}{2} \cdot \cos \frac{\alpha - \beta - 2\varphi}{2} = \sin(\alpha - \beta - 2\varphi).$$

    Thus, $\varphi < \alpha - \beta - 2\varphi$, i.e., $\varphi < \dfrac{\alpha - \beta}{3}$.

5. By the cosine rule, $\cos A = \dfrac{b^2 + c^2 - a^2}{2bc}$. Hence, by the half-angle formulas we have

    $$\sin^2 \frac{A}{2} = \frac{1 - \cos A}{2} = \frac{1}{2} - \frac{b^2 + c^2 - a^2}{4bc} = \frac{a^2 - (b^2 + c^2 - 2bc)}{4bc}$$

    $$= \frac{a^2 - (b-c)^2}{4bc} = \frac{(a - b + c)(a + b - c)}{4bc}$$

    $$= \frac{(2s - 2b)(2s - 2c)}{4bc} = \frac{(s - b)(s - c)}{bc},$$

where $2s = a + b + c$ is the perimeter of triangle ABC. It follows that

$$\sin \frac{A}{2} = \sqrt{\frac{(s-b)(s-c)}{bc}},$$

and the analogous formulas for $\sin \frac{B}{2}$ and $\sin \frac{C}{2}$. Hence

$$\sin \frac{A}{2} \sin \frac{B}{2} \sin \frac{C}{2} = \frac{(s-a)(s-b)(s-c)}{abc} = \frac{[ABC]^2}{sabc}$$

by Heron's formula. It follows that

$$\sin \frac{A}{2} \sin \frac{B}{2} \sin \frac{C}{2} = \frac{[ABC]}{s} \cdot \frac{[ABC]}{abc} = r \cdot \frac{1}{4R},$$

from which (a) follows.

Now we prove (b). By the extended sine rule, we have $a \cos A = 2R \sin A \cdot \cos A = R \sin 2A$. Likewise, $b \cos B = R \sin 2B$ and $c \cos C = R \sin 2C$. By (a) and (b) in Q8 of Testing Question (A), we have

$$4R \sin A \sin B \sin C = \frac{abc}{2R^2}.$$

It suffices to show that $\sin 2A + \sin 2B + \sin 2C = 4 \sin A \sin B \sin C$, which is given by

$$\sin 2A + \sin 2B + \sin 2C$$
$$= 2 \sin(A+B) \cos(A-B) + 2 \sin C \cos C$$
$$= 2 \sin(A+B) \cos(A-B) - 2 \sin(A+B) \cos(A+B)$$
$$= 2 \sin(A+B)[\cos(A-B) - \cos(A+B)]$$
$$= 4 \sin(A+B) \sin B \sin A = 4 \sin C \sin B \sin A.$$

Solutions to Test Questions   7

Testing Questions   (7-A)

1. $256 \sin 10° \sin 30° \sin 50° \sin 70° = 256 \cos 20° \cos 40° \cos 60° \cos 80°$
$$= \frac{128 \sin 20° \cos 20° \cos 40° \cos 80°}{\sin 20°}$$
$$= \frac{64 \sin 40° \cos 40° \cos 80°}{\sin 20°} = \frac{32 \sin 80° \cos 80°}{\sin 20°}$$
$$= \frac{16 \sin 160°}{\sin 20°} = 16.$$

2. All the irreducible proper fractions with the denominator 24 are

$$\frac{1}{24}, \frac{5}{24}, \frac{7}{24}, \frac{11}{24}, \frac{13}{24}, \frac{17}{24}, \frac{19}{24}, \frac{23}{24}.$$

Since $\frac{1}{24} + \frac{23}{24} = \frac{5}{24} + \frac{19}{24} = \frac{7}{24} + \frac{17}{24} = \frac{11}{24} + \frac{13}{24} = 1$ and that
$\cos\alpha + \cos(\pi - \alpha) = 0$, it follows that

$$\sum_{i=1}^{n} \cos(a_i \pi) = 0 + 0 + 0 + 0 = 0.$$

3. For any $k = 1, 2, \cdots,$

$$\sin\frac{1}{2}\beta\cos(\alpha + k\beta) = \frac{1}{2}\left[\sin\left(\alpha + \frac{2k+1}{2}\beta\right) - \sin\left(\alpha + \frac{2k-1}{2}\beta\right)\right]$$

implies that

$$\sum_{k=1}^{n} \sin\frac{1}{2}\beta \cdot \cos(\alpha + k\beta)$$
$$= \frac{1}{2}\sum_{k=1}^{n}\left[\sin\left(\alpha + \frac{2k+1}{2}\beta\right) - \sin\left(\alpha + \frac{2k-1}{2}\beta\right)\right]$$
$$= \frac{1}{2}\left[\sin\left(\alpha + \frac{2n+1}{2}\beta\right) - \sin\left(\alpha + \frac{1}{2}\beta\right)\right]$$
$$= \sin\frac{n\beta}{2}\cos\left(\alpha + \frac{n+1}{2}\beta\right).$$

4. When $\alpha + \beta = 45°$, then

$$1 = \tan 45° = \frac{\tan\alpha + \tan\beta}{1 - \tan\alpha\tan\beta} \Rightarrow 1 - \tan\alpha - \tan\beta = \tan\alpha\tan\beta,$$

therefore

$$(\cot\alpha - 1)(\cot\beta - 1) = \frac{(1 - \tan\alpha)(1 - \tan\beta)}{\tan\alpha\tan\beta} = \frac{2\tan\alpha\tan\beta}{\tan\alpha\tan\beta} = 2.$$

Thus,

$$(\cot 25° - 1)(\cot 24° - 1)\cdots(\cot 20° - 1)$$
$$= [(\cot 25° - 1)(\cot 20° - 1)]\cdots[(\cot 23° - 1)(\cot 22° - 1)]$$
$$= 2^3 = 8.$$

5. $\tan 15° = \dfrac{\tan 60° - \tan 45°}{1 + \tan 60° \tan 45°} = \dfrac{\sqrt{3} - 1}{1 + \sqrt{3}} = 2 - \sqrt{3}$, and $\cot 15° = \dfrac{1}{2 - \sqrt{3}} = 2 + \sqrt{3}$, so $\tan^n 15° + \cot^n 15° = (2 - \sqrt{3})^n + (2 + \sqrt{3})^n$.

Using the binomial expansion, it follows that

$$(2 - \sqrt{3})^n = 2^n - \binom{n}{1} 2^{n-1} \sqrt{3} + \binom{n}{2} 2^{n-2} (\sqrt{3})^2 - \cdots + (-\sqrt{3})^n,$$

$$(2 + \sqrt{3})^n = 2^n + \binom{n}{1} 2^{n-1} \sqrt{3} + \binom{n}{2} 2^{n-2} (\sqrt{3})^2 + \cdots + (\sqrt{3})^n.$$

Therefore in the sum $(2 - \sqrt{3})^n + (2 + \sqrt{3})^n$ only the terms with even powers of $\sqrt{3}$ appear, and each of them appeared in pair, so the sum is an even positive integer.

6. The formula $\tan(k + 1)\alpha = \dfrac{\tan k\alpha + \tan \alpha}{1 - \tan \alpha \tan k\alpha}$ gives

$$\tan k\alpha \tan(k + 1)\alpha = \left(\frac{\tan(k + 1)\alpha}{\tan \alpha} - \frac{\tan k\alpha}{\tan \alpha} \right) - 1,$$

therefore

$$\begin{aligned}
\sum_{k=1}^{n-1} \tan k\alpha \tan(k + 1)\alpha &= \sum_{k=1}^{n-1} \left[\frac{\tan(k + 1)\alpha}{\tan \alpha} - \frac{\tan k\alpha}{\tan \alpha} \right] - (n - 1) \\
&= \frac{\tan n\alpha}{\tan \alpha} - n.
\end{aligned}$$

7. Write the given equality in the form

$$(\cos \alpha + \sin \beta + \sqrt{2}) \cos x + (\cos \beta - \sin \alpha) \sin x = 0,$$

which holds for any real x, so

$$\begin{cases} \cos \alpha + \sin \beta + \sqrt{2} = 0, \\ \cos \beta - \sin \alpha = 0, \end{cases} \quad \text{or} \quad \begin{cases} \sin \beta = -\cos \alpha - \sqrt{2}, \\ \cos \beta = \sin \alpha. \end{cases}$$

By taking squares to both sides of each equality and add up them, then

$$(-\cos \alpha - \sqrt{2})^2 + \sin^2 \alpha = 1$$

which gives the solution $\cos \alpha = -\dfrac{1}{\sqrt{2}}$. Further, $0 < \alpha < \pi$ implies that

$$\alpha = \frac{3\pi}{4},$$

so $\cos \beta = \sin \alpha = \dfrac{1}{\sqrt{2}}$. Thus, $\beta = \dfrac{7\pi}{4}$ since $\pi < \beta < 2\pi$.

8. By using the formula in Q3 of Testing Question (A) (for $\alpha = \beta = 2, n = 180$),

$$\begin{aligned}
&\sin^2 1° + \sin^2 2° + \sin^2 3° + \cdots + \sin^2 360° \\
&= 2(\sin^2 1° + \sin^2 2° + \sin^2 3° + \cdots + \sin^2 180°) \\
&= 180 - (\cos 2° + \cos 4° + \cos 6° + \cdots + \cos 360°) \\
&= 180 - \sin 180° \cos 181° / \sin 1° = 180.
\end{aligned}$$

9. Note that

$$\sin 1° = \sin[(x+1)° - x°] = \sin(x+1)° \cos x° - \cos(x+1)° \sin x°.$$

Thus

$$\begin{aligned}
\frac{\sin 1°}{\sin x° \sin(x+1)°} &= \frac{\cos x° \sin(x+1)° - \sin x° \cos(x+1)°}{\sin x° \sin(x+1)°} \\
&= \cot x° - \cot(x+1)°.
\end{aligned}$$

Multiplying both sides of the given equation by $\sin 1°$, we have

$$\begin{aligned}
\frac{\sin 1°}{\sin n°} &= (\cot 45° - \cot 46°) + (\cot 47° - \cot 48°) + \cdots \\
&\quad + (\cot 133° - \cot 134°) \\
&= \cot 45° - (\cot 46° + \cot 134°) + (\cot 47° + \cot 133°) - \cdots \\
&\quad + (\cot 89° + \cot 91°) - \cot 90° \\
&= 1.
\end{aligned}$$

Therefore, $\sin n° = \sin 1°$, and the least possible integer value for n is 1.

10. $(\sin \alpha + \sin \beta)^2 = \dfrac{2}{3}$, $(\cos \alpha + \cos \beta)^2 = \dfrac{1}{3}$. By adding up them, it is obtained that

$$2 + 2(\sin \alpha \cdot \sin \beta + \cos \alpha \cdot \cos \beta) = 1,$$

so $1 + \cos(\alpha - \beta) = \dfrac{1}{2}$, hence $\cos^2 \dfrac{\alpha - \beta}{2} = \dfrac{1}{4}$.

Testing Questions (7-B)

1. From the formulas for changing sum or difference to product,

$$\cos\frac{\pi}{15} - \cos\frac{2\pi}{15} - \cos\frac{4\pi}{15} + \cos\frac{7\pi}{15}$$

$$= \left(\cos\frac{\pi}{15} + \cos\frac{7\pi}{15}\right) - \left(\cos\frac{2\pi}{15} + \cos\frac{4\pi}{15}\right)$$

$$= 2\cos\frac{4\pi}{15}\cos\frac{\pi}{5} - 2\cos\frac{\pi}{5}\cos\frac{\pi}{15} = 2\cos\frac{\pi}{5}\left(\cos\frac{4\pi}{15} - \cos\frac{\pi}{15}\right)$$

$$= -4\cos\frac{\pi}{5}\sin\frac{\pi}{10}\sin\frac{\pi}{6} = -2\cos\frac{\pi}{5}\sin\frac{\pi}{10} = -\frac{1}{2},$$

since $\cos 18° = \sin 72° = 2\sin 36° \cos 36° = 4\sin 18° \cos 18° \cos 36°$ implies that

$$1 = 4\sin 18° \cos 36°, \quad \therefore\ 2\sin 18° \cos 36° = \frac{1}{2}.$$

2. Note that

$$\cos 36° - \cos 72° = \frac{2(\cos 36° - \cos 72°)(\cos 36° + \cos 72°)}{2(\cos 36° + \cos 72°)}$$

$$= \frac{2\cos^2 36° - 2\cos^2 72°}{2(\cos 36° + \cos 72°)}.$$

By the double-angle formulas, the above equality becomes

$$\cos 36° - \cos 72° = \frac{\cos 72° + 1 - \cos 144° - 1}{2(\cos 36° + \cos 72°)}$$

$$= \frac{\cos 72° + \cos 36°}{2(\cos 36° + \cos 72°)} = \frac{1}{2}.$$

3. By using the double angle formulas and the half angle formulas,

$$\frac{\cos 100°}{1 - 4\sin 25° \cos 25° \cos 50°} = \frac{\cos 100°}{1 - 2\sin 50 \cos 50°}$$

$$= \frac{\cos^2 50° - \sin^2 50°}{(\cos 50° - \sin 50°)^2} = \frac{\cos 50° + \sin 50°}{\cos 50° - \sin 50°}$$

$$= \frac{1 + \tan 50°}{1 - \tan 50°} = \frac{\tan 45° + \tan 50°}{1 - \tan 45° \tan 50°} = \tan 95°, \therefore x = 95.$$

4. The left-hand side of the desired equation is equal to

$$\sum_{k=1}^{89} \frac{1}{\sin k° \sin(k+1)°} = \frac{1}{\sin 1°} \sum_{k=1}^{89} [\cot k° - \cot(k+1)°]$$

$$= \frac{1}{\sin 1°} \cdot \cot 1° = \frac{\cos 1°}{\sin^2 1°},$$

5. We construct an equation with roots $\tan \dfrac{n\pi}{5}$, $n = 0, 1, 2, 3, 4$ as follows.

Since the equation $\tan 5\theta = 0$ for $\theta \in [0, \pi)$ has roots $\dfrac{n\pi}{5}$, $n = 0, 1, 2, 3, 4$, then each of the five roots satisfies the equation $\tan 3\theta = -\tan 2\theta$, therefore, by the multiple angle formulae, it satisfies the equation

$$\frac{3\tan\theta - \tan^3\theta}{1 - 3\tan^2\theta} = \frac{-2\tan\theta}{1 - \tan^2\theta}.$$

Letting $x = \tan\theta$, we have $\dfrac{3x - x^3}{1 - 3x^2} = \dfrac{-2x}{1 - x^2}$, or, equivalently,

$$x(x^4 - 10x^2 + 5) = 0.$$

If consider non-zero roots, then it becomes

$$x^4 - 10x^2 + 5 = 0. \tag{15.33}$$

Thus, $\tan\theta$ for $\theta = \dfrac{n\pi}{5}$, $n = 1, 2, 3, 4$ are the four roots of (15.33). By the Viete's theorem,

$$\tan\frac{\pi}{5} \cdot \tan\frac{2\pi}{5} \cdot \tan\frac{3\pi}{5} \cdot \tan\frac{4\pi}{5} = 5, \tag{15.34}$$

$$\tan\frac{\pi}{5} \cdot \tan\frac{2\pi}{5} + \tan\frac{\pi}{5} \cdot \tan\frac{3\pi}{5} + \tan\frac{\pi}{5} \cdot \tan\frac{4\pi}{5} + \tan\frac{2\pi}{5} \cdot \tan\frac{3\pi}{5}$$
$$+ \tan\frac{2\pi}{5} \cdot \tan\frac{4\pi}{5} + \tan\frac{3\pi}{5} \cdot \tan\frac{4\pi}{5} = -10.$$

Since $\tan\dfrac{\pi}{5} > 0$, $\tan\dfrac{2\pi}{5} > 0$ and $\tan\dfrac{3\pi}{5} = -\tan\dfrac{2\pi}{5}$, $\tan\dfrac{4\pi}{5} = -\tan\dfrac{\pi}{5}$, (15.34) drives

$$\tan^2\frac{\pi}{5} \cdot \tan^2\frac{2\pi}{5} = 5, \quad \therefore \tan\frac{\pi}{5} \cdot \tan\frac{2\pi}{5} = \sqrt{5}, \text{ (i) is proven.}$$

From (15.34),

$$\tan\frac{\pi}{5}\cdot\tan\frac{2\pi}{5} - \tan\frac{\pi}{5}\cdot\tan\frac{2\pi}{5} - \tan^2\frac{\pi}{5} - \tan^2\frac{2\pi}{5}$$

$$-\tan\frac{2\pi}{5}\cdot\tan\frac{\pi}{5} + \tan\frac{2\pi}{5}\cdot\tan\frac{\pi}{5} = -10,$$

$$\therefore \tan^2\frac{\pi}{5} + \tan^2\frac{2\pi}{5} = 10, \text{ (ii) is proven.}$$

Solutions to Testing Questions 8

Testing Question (8-A)

1. For any given real number a, the graph of f is a polygonal line. Therefore the minimum value $\frac{3}{2}$ is taken when the graph is at some turning point, i.e., $x + 1 = 0$ or $ax + 1 = 0$.

   When $a = 0$, then $f(x) = |x + 1| + 1 \geq 1$ and $f(-1) = 1$, i.e. the minimum value of f is 1. Therefore $a \neq 0$.

   (i) When $x = -1$, then $f(-1) = |1 - a|$. $f(-1) = \frac{3}{2} \Rightarrow a = -\frac{1}{2}$ or $a = \frac{5}{2}$.

   $$a = -\frac{1}{2} \Rightarrow f(x) = |x+1| + \left|-\frac{1}{2}x + 1\right| = \begin{cases} -\frac{3}{2}x, & x \leq -1, \\ 2 + \frac{1}{2}x, & -1 < x \leq 2, \\ \frac{3}{2}x, & 2 < x. \end{cases}$$

   therefore $f(x) \geq \frac{3}{2}$ for $x \in \mathbb{R}$.

   For $a = \frac{5}{2}$, $f\left(-\frac{1}{2}\right) = \frac{3}{4} < \frac{3}{2}$, so only $a = -\frac{1}{2}$ satisfies the requirement.

   (ii) When $ax + 1 = 0$ i.e. $x = -\frac{1}{a}$, then $\left|-\frac{1}{a} + 1\right| = \frac{3}{2} \Rightarrow a = -2$ or $a = \frac{2}{5}$.

   $$a = -2 \Rightarrow f(x) = |x + 1| + |-2x + 1| = \begin{cases} -3x, & x \leq -1, \\ 2 - x, & -1 < x \leq \frac{1}{2}, \\ 3x, & \frac{1}{2} < x. \end{cases}$$

therefore $f(x) \geq \dfrac{3}{2}$ for $x \in \mathbb{R}$.

For $a = \dfrac{2}{5}$, $f\left(-\dfrac{3}{2}\right) = \dfrac{9}{10} < \dfrac{3}{2}$, so only $a = -2$ satisfies the requirement.

Thus, $a = -\dfrac{1}{2}$ or $a = -2$.

2. Let $u = x^2 + 4x$, then $u = (x + 2)^2 - 4 \geq -4$ and

$$y = (u + 5)(u + 1) + 3u + 5 = u^2 + 9u + 10 = (u + \dfrac{9}{2})^2 - \dfrac{41}{4}.$$

Since $u \geq -4$, so the minimum value of y is taken when $u = -4$ (i.e. $x = -2$), therefore $\quad y_{\min} = (-4)^2 + 9(-4) + 10 = -10$.

3. $y = \dfrac{x^2 + n}{x^2 + x + 1} \Rightarrow (y - 1)x^2 + yx + (y - n) = 0$.

When $y = 1$, then $x = n - 1$, so 1 is in the range of y.

When $y \neq 1$, then above quadratic equation in x has real roots for any $y \neq 1$ in its range. So its discriminant is non-negative. Therefore

$$\Delta = y^2 - 4(y - 1)(y - n) \geq 0 \Rightarrow 3y^2 - 4(n + 1)y + 4n \leq 0$$

$$\Rightarrow a_n + b_n = \dfrac{4(n + 1)}{3}, \quad a_n b_n = \dfrac{4n}{3},$$

therefore $(a_n - b_n)^2 = (a_n + b_n)^2 - 4a_n b_n = \dfrac{16(n + 1)^2 - 48n}{9} = \dfrac{16(n^2 - n + 1)}{9}$, so $a_n - b_n = \dfrac{4}{3}\sqrt{n^2 - n + 1}$.

4. As shown in the graph of y below, it is easy to find the coordinates of the points in it are

$$E(-\sqrt{2}, 0), \quad F(\sqrt{2}, 0), \quad D(0, 1).$$

Since

$$y = 1 - \dfrac{x^2}{2} \text{ if } |x| \leq \sqrt{2}$$

and

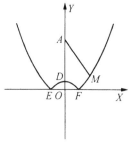

$$y = \dfrac{x^2}{2} - 1 \text{ when } |x| \geq \sqrt{2},$$

$|AM|_{\min} = AD = a - 1$ when $|x| \leq \sqrt{2}$, and

$$|AM|^2 = x^2 + \left(\dfrac{x}{2} - 1 - a\right)^2 = \dfrac{1}{4}(x^2 - 2a)^2 + 2a + 1$$

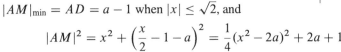

when $|x| \geq \sqrt{2}$, which takes its minimum value $2a + 1$ when $x = \sqrt{2a} > \sqrt{2}$, so $|AM|_{\min} = \sqrt{2a + 1}$ at $x = \sqrt{2a}$.

Now $|AD|^2 - (2a+1) = (a-1)^2 - (2a+1) = a(a-4)$, so $AD < \sqrt{2a + 1}$ when $1 < a < 4$.

When $a > 4$, then $\sqrt{2a + 1} < a - 1$. Thus,

$$|AM|_{\min} = a - 1 \text{ if } 1 < a \leq 4, \quad \text{and} \quad |AM|_{\min} = \sqrt{2a + 1} \text{ if } a > 4.$$

5. $\dfrac{a^2}{x} + \dfrac{b^2}{1 - x} = a^2 + b^2 + a^2\left(\dfrac{1}{x} - 1\right) + b^2\left(\dfrac{1}{1 - x} - 1\right)$. By the mean inequality,

$$\frac{a^2}{x} + \frac{b^2}{1 - x} \geq a^2 + b^2 + 2\sqrt{a^2 b^2 \frac{1 - x}{x} \cdot \frac{x}{1 - x}} = (a + b)^2.$$

The value $(a + b)^2$ is obtainable when $\dfrac{a^2(1 - x)}{x} = \dfrac{b^2 x}{1 - x}$, namely $x = \dfrac{a}{a + b}$. Thus, the minimum value of f is $(a + b)^2$.

6. Since $2 - x > 0$, by the mean inequality,

$$f(x) = \frac{1 + (2 - x)^2}{2 - x} = \frac{1}{2 - x} + (2 - x) \geq 2\sqrt{\frac{1}{2 - x} \cdot (2 - x)} = 2.$$

the equality holds when $\dfrac{1}{2 - x} = 2 - x$, i.e., $x = 1$, so 2 is reachable by f. Thus, $f(x)_{min} = 2$.

7. (i) By completing the squares, the given equality yields

$$\left(x + \frac{1}{2}\right)^2 + (y + 1)^2 + \left(z + \frac{3}{2}\right)^2 = \frac{27}{4}.$$

Similar to Example 5, the following inequality holds for any $a, b, c \in \mathbb{R}$:

$$3(a^2 + b^2 + c^2) \geq (a + b + c)^2,$$

and the equality holds if and only if $a = b = c$. Therefore

$$\left[\left(x + \frac{1}{2}\right) + (y + 1) + \left(z + \frac{3}{2}\right)\right]^2$$

$$\leq 3\left[\left(x + \frac{1}{2}\right)^2 + (y + 1)^2 + \left(z + \frac{3}{2}\right)^2\right] = \frac{81}{4}.$$

Thus, $x + y + z \leq \dfrac{9}{2} - 3 = \dfrac{3}{2}$, the equality holds if and only if $x = 1$, $y = \dfrac{1}{2}$, $z = 0$.

(ii) $(x + y + z)^2 \geq x^2 + y^2 + z^2$, $3(x + y + z) \geq x + 2y + 3z$ implies that

$$\frac{13}{4} = x^2 + y^2 + z^2 + z + 2y + 3z \leq (x + y + z)^2 + 3(x + y + z),$$

therefore

$$x + y + z \geq \frac{\sqrt{22} - 3}{2}.$$

The equality holds if and only if $x = 0$, $y = 0$, $z = \dfrac{\sqrt{22} - 3}{2}$.

8.   Let the lengths of edges PA, PB, PC be a, b, c respectively.

Then volume of the tetrahedron is $V = \dfrac{1}{6}abc$.

By assumption,

$$S = a + b + c + \sqrt{a^2 + b^2} + \sqrt{b^2 + c^2} + \sqrt{c^2 + a^2}. \tag{15.35}$$

By Mean Inequality,

$$a + b + c \geq 3(abc)^{\frac{1}{3}} \tag{15.36}$$

and

$$\sqrt{a^2 + b^2} + \sqrt{b^2 + c^2} + \sqrt{c^2 + a^2}$$
$$\geq \sqrt{2ab} + \sqrt{2bc} + \sqrt{2ca} \geq 3\sqrt{2}(abc)^{\frac{1}{3}}, \tag{15.37}$$

where the equalities hold when and only when $a = b = c$. Substituting (15.36) and (15.37) into (15.35), it is obtained that

$$S \geq 3(1 + \sqrt{2})(abc)^{\frac{1}{3}},$$

therefore

$$V = \frac{1}{6}abc \leq \frac{S^3}{162(1 + \sqrt{2})^3},$$

where the equalities hold when and only when $a = b = c$. Thus,

$$V_{\max} = \frac{S^3}{162(1 + \sqrt{2})^3}.$$

9. $2^a + 4^b = 2^{17-b} + 2^{2b} = 2^{16-b} + 2^{16-b} + 2^{2b} \geq 3\sqrt[3]{2^{16-b}2^{16-b}2^{2b}}$
    $= 3 \cdot 2^{\frac{32}{3}} = 3072\sqrt[3]{4} \left(\text{when } b = \frac{16}{3} \right).$

10. As shown in the diagram (1) below, let $ABCD$ be a convex quadrilateral with $AB = a$, $CD = b$ and $BD = c$, where $a + b + c = 16\sqrt{2}$ cm, and the area of $ABCD$ is 64 cm². From

$$64 = \frac{1}{2}ac \sin \angle ABD + \frac{1}{2}bc \sin \angle CDB \leq \frac{1}{2}(a+b)c \leq \frac{1}{8}(a+b+c)^2 = 64,$$

the two equal signs hold, therefore $a + b = c = 8\sqrt{2}$ and

$$AB \perp BD \qquad \text{and} \qquad CD \perp BD,$$

as shown in the diagram (2). Hence,

$$AC = \sqrt{(a+b)^2 + c^2} = \sqrt{2(8\sqrt{2})^2} = 16.$$

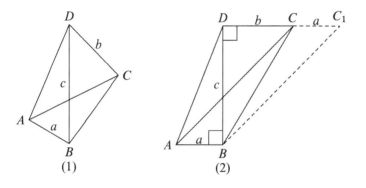

(1)   (2)

Testing Questions   (8-B)

1. The given conditions gives

$$f(0) = -|1 + a| + 2 \geq 0 \quad \text{and} \quad f(1) = -|a| + 2 \geq 0,$$

from which $-2 \leq a \leq 1$ is obtained. Now for $a = -2$,

$$f(x) = x^2 - 2x - |x + 1| - |x - 2| + 4$$

$$= \begin{cases} x^2 + 3 & x < -1, \\ x^2 - 2x + 1 & -1 \leq x \leq 2, \\ x^2 - 4x + 5 & x > 2. \end{cases}$$

Thus, $f(x) \geq 0$ for all $x \in \mathbb{R}$, so $a_{\min} = -2$.

2. Since $a > 0$, $f(x)$ takes its minimum value when $x = x_0 = \dfrac{a+b}{3a}$, and

$$|f(x_0)| = \left| \frac{(3a)b - (a+b)^2}{(3a)} \right| = \frac{a^2 + b^2 - ab}{3a}.$$

(I) When $0 \leq x_0 \leq 1$, then $a + b \geq 0$ and $2a - b \geq 0$.
   (i) If $b \leq 0$, then

$$|f(1)| = a - b = f(1) > -b = |f(0)|; \quad \text{and}$$
$$|f(x_0)| \leq |f(1)| \Leftrightarrow a^2 + b^2 - ab \leq 3a(a-b) \Leftrightarrow (a+b)^2 \leq 3a^2$$
$$\Leftrightarrow 0 \leq a + b \leq \sqrt{3}a \text{ (which is true obviously)}.$$

Therefore $\max\limits_{0 \leq x \leq 1} |f(x)| = f(1)$, the conclusion is true in the case.

   (ii) When $0 < b \leq \dfrac{a}{2}$, then $a \geq 2b > 0$, so

$$|f(1)| = a - b = f(1) \geq b = f(0) = |f(0)|$$

and similar to above, $|f(x_0)| \leq f(1)$, so the conclusion is true also.

   (iii) When $b > \dfrac{a}{2} > 0$, then $0 < a < 2b$, so

$$|f(1)| = |a - b| < b = f(0),$$
$$|f(x_0)| \leq f(0) \Leftrightarrow a^2 + b^2 - ab \leq 3ab \Leftrightarrow (2a - b)^2 \leq 3a^2$$
$$\Leftrightarrow 2a - b \leq \sqrt{3}a \Leftrightarrow (2 - \sqrt{3})a < b,$$

and the last inequality is true since $(2 - \sqrt{3})a < \dfrac{1}{2}a < b$. Thus, the conclusion is true also.

(II) When $x_0 < 0$, then $a + b < 0$, so $0 > b = f(0)$. Thus,

$$\max\limits_{0 \leq x \leq 1} |f(x)| = \max\{|f(0)|, |f(1)|\} = \max\{-b, a - b\}$$
$$= a - b = f(1) = \max\{f(0), f(1)\}.$$

(III) When $x_0 > 1$, then $2a - b < 0$, so $b > 0$. Thus,

$$\max\limits_{0 \leq x \leq 1} |f(x)| = \max\{|f(0)|, |f(1)|\} = \max\{b, |a - b|\}$$
$$= \max\{b, b - a\} = b = f(0) = \max\{f(0), f(1)\}.$$

In summary, we have proven that $\max\limits_{0 \leq x \leq 1} |f(x)| \leq \max\{f(0), f(1)\}$.

3. Since $\displaystyle\sum_{i=1}^{2010} \frac{x_i^{2008}}{1 - x_i^{2009}} = \sum_{i=1}^{2010} \frac{x_i^{2009}}{x_i(1 - x_i^{2009})}$, let $y_i = x_i(1 - x_i^{2009})$, then for any $1 \le i \le 2010$,

$$
\begin{aligned}
y_i^{2009} &= \frac{1}{2009}(2009x_i^{2009})(1 - x_i^{2009})^{2009} \\
&\le \frac{1}{2009}\left(\frac{2009x_i^{2009} + 2009(1 - x_i^{2009})}{2010}\right)^{2010} \\
&= \frac{1}{2009}\left(\frac{2009}{2010}\right)^{2010} = (2009)^{2009}\left(\frac{1}{2010}\right)^{2010},
\end{aligned}
$$

therefore

$$
y_i \le 2009(2010)^{-\frac{2010}{2009}} \Rightarrow \frac{1}{y_i} \ge \frac{(2010)^{\frac{2010}{2009}}}{2009} = \frac{2010}{2009} \cdot \sqrt[2009]{2010}.
$$

Thus,

$$
\sum_{i=1}^{2010} \frac{x_i^{2008}}{1 - x_i^{2009}} = \sum_{i=1}^{2010} \frac{x_i^{2009}}{y_i} \ge \frac{2010}{2009} \cdot \sqrt[2009]{2010}.
$$

The equality holds when $2009x_i^{2009} = 1 - x_i^{2009}$, namely $x_i = \dfrac{1}{\sqrt[2009]{2010}}$ for $i = 1, 2, \ldots, 2010$.

Thus, $\displaystyle\min\left\{\sum_{i=1}^{2010} \frac{x_i^{2008}}{1 - x_i^{2009}}\right\} = \frac{2010}{2009} \cdot \sqrt[2009]{2010}.$

4. If we interpret x_1 and x_2 are the coordinates of a point; that is, assume that $P = (x_1, x_2)$, then P lies on a circle centered at the origin with radius c. We can describe the circle parametrically; that is, write $x_1 = c \cos \theta, x_2 = c \sin \theta$, and similarly, $y_1 = c \cos \varphi, y_2 = c \sin \varphi$. Then

$$
\begin{aligned}
S &= 2 - c(\cos \theta + \sin \theta + \cos \varphi + \sin \varphi) + c^2(\cos \theta \cos \varphi + \sin \theta \sin \varphi) \\
&= 2 - \sqrt{2}c[\sin(\theta + \tfrac{\pi}{4}) + \sin(\varphi + \tfrac{\pi}{4})] + c^2 \cos(\theta - \varphi) \\
&\le 2 + 2\sqrt{2}c + c^2 = (\sqrt{2} + c)^2,
\end{aligned}
$$

where the equality holds when $\theta = \varphi = 5\pi/4$, that is, $x_1 = x_2 = y_1 = y_2 = -\dfrac{\sqrt{2}}{2}c$.

5. Let $f(t) = t + \frac{1}{t}, t > 0$ and $g(x) = \max\{f(ax), f(bx)\}, x > 0$. The problem is converted to finding the minimum value of g on $(0, +\infty)$.

When $a = b$, then $g(x) = ax + \dfrac{1}{ax} \geq 2$, and the equality holds when $ax = 1$.

When $0 < a < b$, then

$$f(bx) - f(ax) = bx - ax + \frac{1}{bx} - \frac{1}{ax} = (b-a)x \left(1 - \frac{1}{abx^2} \right),$$

therefore

$$g(x) = \begin{cases} f(bx) & \text{if } x \geq \frac{1}{\sqrt{ab}}, \\ f(ax) & \text{if } x < \frac{1}{\sqrt{ab}}. \end{cases}$$

Since $f(t) = t + \dfrac{1}{t}, t > 0$ is decreasing on $(0, 1]$ and increasing on $[1, +\infty)$,

if $x \geq \dfrac{1}{\sqrt{ab}}$, then $bx \geq \sqrt{\dfrac{b}{a}} > 1$, so $f(bx) \geq f\left(\sqrt{\dfrac{b}{a}} \right)$, i.e.

$$c = f\left(\sqrt{\frac{b}{a}} \right) = \sqrt{\frac{b}{a}} + \sqrt{\frac{a}{b}},$$

and the equality holds when $x = \dfrac{1}{\sqrt{ab}}$.

If $0 < x \leq \dfrac{1}{\sqrt{ab}}$, then $ax \leq \sqrt{\dfrac{a}{b}} < 1$, so $f(ax) \geq f\left(\sqrt{\dfrac{a}{b}} \right)$, i.e.

$$c = f\left(\sqrt{\frac{a}{b}} \right) = f\left(\sqrt{\frac{b}{a}} \right) = \sqrt{\frac{b}{a}} + \sqrt{\frac{a}{b}},$$

and the equality holds when $x = \dfrac{1}{\sqrt{ab}}$.

Thus, $c_{\max} = \sqrt{\dfrac{b}{a}} + \sqrt{\dfrac{a}{b}}$.

Solutions to Testing Questions   9

Testing Questions   (9-A)

1. From

$$\sin(x + \alpha) + \sin(x - \alpha) - 4\sin x = 2\sin x(\cos\alpha - 2),$$
$$\cos(x + \alpha) + \cos(x - \alpha) - 4\cos x = 2\cos x(\cos\alpha - 2),$$

and $\cos\alpha - 2 \neq 0$, it follows that $f(x) = \tan x$. $\tan x$ is increasing in $\left(0, \frac{\pi}{4}\right)$, so the range of f is $(0, 1)$.

2. The half angle formula and the R-formula give

$$
\begin{aligned}
f(x) &= \frac{1}{2}(1 + \cos 2\theta x) + \frac{1}{2}\sin 2\theta x = \frac{1}{2} + \frac{1}{2}(\cos 2\theta x + \sin 2\theta x) \\
&= \frac{1}{2} + \frac{\sqrt{2}}{2}\sin\left(2\theta x + \frac{\pi}{4}\right).
\end{aligned}
$$

Since f has the minimum period $\frac{\pi}{2}$, so $\frac{2\pi}{2\theta} = \frac{\pi}{2}$, i.e. $\theta = 2$. Thus,

$$\theta f(x) = 1 + \sqrt{2}\sin\left(4x + \frac{\pi}{4}\right) \Rightarrow f_{\max} = 1 + \sqrt{2}.$$

3. Let $\sin x = t$, then $-1 \leq t \leq 1$ and $y = g(t) = (-at^2 + a - 3)t$. Then

$$
\begin{aligned}
&-at^3 + (a - 3)t \geq -3 \Leftrightarrow -at(t^2 - 1) - 3(t - 1) \geq 0 \\
&\Leftrightarrow (t - 1)[-at(t + 1) - 3] \geq 0.
\end{aligned}
$$

Since $t - 1 \leq 0$, so $-at(t + 1) - 3 \leq 0$, or

$$at(t + 1) \geq -3, \quad \text{for all} -1 \leq t \leq 1. \qquad (*)$$

$(*)$ holds obviously when $t = 0$ or -1.

$a \geq \dfrac{-3}{t^2 + t} \geq -\dfrac{3}{2}$ when $0 < t \leq 1$.

For $-1 < t < 0$, then $-\dfrac{1}{4} \leq t^2 + t < 0 \Rightarrow a \leq 12$. Thus, $-\dfrac{3}{2} \leq a \leq 12$.

4. Based on $a^2 + b^2 = c^2$ let $a = c\cos\theta, b = c\sin\theta$, where $0 < \theta < \dfrac{\pi}{2}$.

Then

$$
\begin{aligned}
y &= \frac{c^3 \cos^3 \theta + c^3 \sin^3 \theta + c^3}{c(c \cos \theta + c \sin \theta + c)^2} = \frac{\cos^3 \theta + \sin^3 \theta + 1}{(\cos \theta + \sin \theta + 1)^2} \\
&= \frac{(\cos \theta + \sin \theta)(\cos^2 \theta - \cos \theta \sin \theta + \sin^2 \theta) + 1}{(\cos \theta + \sin \theta + 1)^2} \\
&= \frac{(\cos \theta + \sin \theta)(1 - \cos \theta \sin \theta) + 1}{(\cos \theta + \sin \theta + 1)^2}.
\end{aligned}
$$

Let $x = \cos \theta + \sin \theta = \sqrt{2} \sin \left(\theta + \dfrac{\pi}{4} \right)$, then $1 < x \le \sqrt{2}$, so

$$
\begin{aligned}
y &= \frac{x \cdot \left[1 - \frac{1}{2}(x^2 - 1) \right] + 1}{(x + 1)^2} = \frac{2 + 3x - x^3}{2(x + 1)^2} \\
&= \frac{(x + 1)(2 + x - x^2)}{2(x + 1)^2} = \frac{2 + x - x^2}{2(1 + x)} \\
&= \frac{(2 - x)(1 + x)}{2(1 + x)} = \frac{2 - x}{2} = 1 - \frac{1}{2} \cdot x.
\end{aligned}
$$

Since y, as a function of x, is decreasing on $(1, \sqrt{2}]$, so the range of y is $\left[1 - \dfrac{\sqrt{2}}{2}, \dfrac{1}{2} \right)$.

5.
$$
\begin{aligned}
\sum_{1 \le i < j \le n} \cos^2(\alpha_i - \alpha_j) &= \frac{1}{2} \sum_{1 \le i < j \le n} [1 + \cos(2\alpha_i - 2\alpha_j)] \\
&= \frac{n^2 - n}{4} + \frac{1}{2} \sum_{1 \le i < j \le n} (\cos 2\alpha_i \cos 2\alpha_j + \sin 2\alpha_i \sin 2\alpha_j) \\
&= \frac{n^2 - n}{4} + \frac{1}{4} \left[\left(\sum_{i=1}^{n} \cos 2\alpha_i \right)^2 + \left(\sum_{i=1}^{n} \sin 2\alpha_i \right)^2 - n \right] \\
&\ge \frac{n^2 - 2n}{4}.
\end{aligned}
$$

When $\alpha_i = \dfrac{i\pi}{n}$ for all i, then $\displaystyle\sum_{i=1}^{n} \cos 2\alpha_i = \sum_{i=1}^{n} \sin 2\alpha_i = 0$, so the value

$\dfrac{n^2 - 2n}{4}$ is reachable. Thus,

$$\min\left\{\sum_{1 \le i < j \le n} \cos^2(\alpha_i - \alpha_j)\right\} = \frac{n^2 - 2n}{4}.$$

6.   Since $\cos A \ge 0 \Rightarrow (b - c)^2 \cos A \ge 0$, it follows that

$$b^2 + c^2 - 2bc \cos A - (b^2 + c^2)(1 - \cos A) \ge 0.$$

Then the cosine rule gives $a^2 - (b^2 + c^2)(1 - \cos A) \ge 0$, therefore

$$\frac{a^2}{b^2 + b^2} \ge 1 - \cos A = 2 \sin^2 \frac{A}{2}, \quad \text{or} \quad \frac{a}{\sqrt{b^2 + c^2}} \ge \sqrt{2} \sin \frac{A}{2}.$$

Similarly, we have

$$\frac{b}{\sqrt{c^2 + a^2}} \ge \sqrt{2} \sin \frac{B}{2} \quad \text{and} \quad \frac{c}{\sqrt{a^2 + b^2}} \ge \sqrt{2} \sin \frac{C}{2}.$$

Thus,

$$\frac{abc}{\sqrt{(a^2 + b^2)(b^2 + c^2)(c^2 + a^2)}} \ge 2\sqrt{2} \sin \frac{A}{2} \sin \frac{B}{2} \sin \frac{C}{2}.$$

Since

$$2R^2 \sin A \sin B \sin C = \frac{1}{2}(2R \sin A)(2R \sin B) \sin C = [ABC]$$

and $r(a + b + c) = 2[ABC]$ or $2rR = \dfrac{2[ABC]}{\sin A + \sin B + \sin C}$, so

$$\begin{aligned}
\frac{r}{R} &= \frac{2 \sin A \sin B \sin C}{\sin A + \sin B + \sin C} = \frac{2 \sin A \sin B \sin C}{4 \cos \frac{A}{2} \cos \frac{B}{2} \cos \frac{C}{2}} \\
&= 4 \sin \frac{A}{2} \sin \frac{B}{2} \sin \frac{C}{2},
\end{aligned}$$

so

$$\frac{abc}{\sqrt{2(a^2 + b^2)(b^2 + c^2)(c^2 + a^2)}} \ge 2 \sin \frac{A}{2} \sin \frac{B}{2} \sin \frac{C}{2} = \frac{r}{2R}.$$

7. Let $x = \tan\dfrac{\alpha}{2}$, $y = \tan\dfrac{\beta}{2}$, then $x, y \in (0, 1)$ and

$$\cot\alpha + \cot\beta = \frac{1-x^2}{2x} + \frac{1-y^2}{2y} = \frac{(x+y)(1-xy)}{2xy},$$

hence

$$\begin{aligned} A &= \frac{2xy(1-\sqrt{xy})^2}{(x+y)(1-xy)} = \frac{2xy(1-\sqrt{xy})}{(x+y)(1+\sqrt{xy})} \\ &\le \frac{2xy(1-\sqrt{xy})}{2\sqrt{xy}(1+\sqrt{xy})} = \frac{\sqrt{xy}(1-\sqrt{xy})}{1+\sqrt{xy}}. \end{aligned}$$

Let $t = \sqrt{xy}$, then $t \in (0, 1)$ and

$$A = \frac{t(1-t)}{1+t} = \frac{-(1+t)^2 + 3(1+t) - 2}{1+t} = 3 - \left(1 + t + \frac{2}{1+t}\right).$$

Since $1 + t + \dfrac{2}{1+t} = \sqrt{2}\left(\dfrac{1+t}{\sqrt{2}} + \dfrac{\sqrt{2}}{1+t}\right)$ takes its minimum value at

$\dfrac{1+t}{\sqrt{2}} = 1$ i.e. $t = \sqrt{2} - 1$, so $A_{\max} = 3 - 2\sqrt{2}$.

8. $\dfrac{\sin^3\alpha}{\cos\alpha} + \dfrac{\cos^3\alpha}{\sin\alpha} = \dfrac{\sin^4\alpha + \cos^4\alpha}{\sin\alpha\cos\alpha} = \dfrac{1 - 2(\sin\alpha\cos\alpha)^2}{\sin\alpha\cos\alpha}.$

Let $t = \sin\alpha\cos\alpha = \dfrac{1}{2}\sin 2\alpha$, where $0 < t \le \dfrac{1}{2}$, then

$$f(t) = \frac{1-2t^2}{t} = \frac{1}{t} - 2t, \qquad 0 < t \le \frac{1}{2},$$

then f is a decreasing function, so $f_{min} = f(\dfrac{1}{2}) = 1$, the answer is (C).

9. As shown in the right graph below, let $\angle CAB = \alpha$, $\angle ABC = \beta$, $\angle ACB = \gamma$.

Then $\angle BOA_1 = \alpha$, so $OA_1 = \cos\alpha$. Similarly, $OB_1 = \cos\beta$, $OC_1 = \cos\gamma$.
By the HM-AM inequality,

$$\begin{aligned} &\frac{1}{OA_1} + \frac{1}{OB_1} + \frac{1}{OC_1} \\ &= \frac{1}{\cos\alpha} + \frac{1}{\cos\beta} + \frac{1}{\cos\gamma} \\ &\ge \frac{9}{\cos\alpha + \cos\beta + \cos\gamma}. \end{aligned}$$

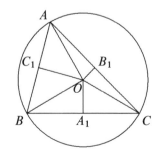

Thus, the conclusion is proven by using the basic inequality (cf. Q3(a) in Appendix C)

$$\cos\alpha + \cos\beta + \cos\gamma \le \frac{3}{2}.$$

10. Let $\sin x - \sin y = t$.

Since $(\cos x + \cos y)^2 + (\sin x - \sin y)^2 = 1 + t^2$,

$$\cos x \cos y - \sin x \sin y = \frac{1}{2}(t^2 - 1),$$

$$\cos(x + y) = \frac{1}{2}(t^2 - 1).$$

Then $-1 \le \cos(x + y) \le 1 \Rightarrow -1 \le \frac{1}{2}(t^2 - 1) \le 1 \Rightarrow 0 \le t^2 \le 3$, so

$$-\sqrt{3} \le t \le \sqrt{3},$$

the answer is (D).

Testing Questions   (9-B)

1. The answer is $n = 4$.

For any triangles T which has interior angles α, β, γ, write $f_n(T) = \sin n\alpha + \sin n\beta + \sin n\gamma$.

Lemma.   If $x + y + z = k\pi$, where $k \in \mathbb{Z}$, then

$$|\sin x| \le |\sin y| + |\sin z|,$$

and the strict inequality holds when there is no $l, m \in \mathbb{Z}$ such that $y = l\pi$ and $z = m\pi$.

Proof of Lemma.   It is easy to see that

$$
\begin{aligned}
|\sin x| &= |\sin(y + z)| = |\sin y \cos z + \cos y \sin z| \\
&\le |\sin y||\cos z| + |\cos y||\sin z| \le |\sin y| + |\sin z|.
\end{aligned}
$$

Further, if there is no $l, m \in \mathbb{Z}$ such that $y = l\pi$ and $z = m\pi$, then $|\cos y| < 1$ and $|\cos z| < 1$, so the strict inequality holds.

In the three terms of $f_n(T)$ there must be at least two with same sign, the lemma implies that the sign of $f_n(T)$ is the same as that of the two terms.

$f_1(T) > 0$ is obvious, and $f_2(T)$ is also positive: In fact, any triangle must have two acute interior angles, say, $0 < \alpha, \beta < \dfrac{\pi}{2}$, then $\sin 2\alpha > 0, \sin 2\beta > 0$.

For $n = 3$, consider an isosceles acute triangle. Let the two base angles be $\alpha = \beta = x$. When x increases from $\dfrac{\pi}{4}$, then $3x$ increases from $\dfrac{3\pi}{4}$ to $\dfrac{3\pi}{2}$, therefore $\sin 3x$ and $f_3(T)$ can take both positive and negative values.

Let $n = 4$, and $\alpha \geq \beta \geq \gamma$. Since $\alpha < \dfrac{\pi}{2}$, so $\beta > \dfrac{\pi}{4}$. Thus, $\pi < 4\beta \leq 4\alpha < 2\pi$, so that $\sin 4\alpha < 0, \sin 4\beta < 0$ which implies that $f_4(T) < 0$.

For the case $n > 4$, consider an isosceles triangle with the base angle $\alpha = \beta = x$. When x increases from $\dfrac{\pi}{4}$ to $\dfrac{\pi}{2}$, since the change of nx is greater than π, so there are x_1, x_2 such that $\sin nx_1 < 0, \sin nx_2 > 0$, therefore $f_n(T)$ may be negative and also may positive.

Thus, $n = 4$ is the unique solution.

2. $|\sin t| \leq 1$ for any $t \in \mathbb{R}$ implies that $\sin^4 t \leq \sin^2 t$, and the equality holds if and only if $\sin^2 t = 0$ or $\sin^2 t = 1$. Therefore, by $ab \leq \dfrac{1}{2}(a^2 + b^2)$ for any real numbers a, b,

$$\sin^2 x \cos y + \sin^2 y \cos z + \sin^2 z \cos x$$

$$\leq \frac{1}{2}(\sin^4 x + \cos^2 y) + \frac{1}{2}(\sin^4 y + \cos^2 z) + \frac{1}{2}(\sin^4 z + \cos^2 x)$$

$$= \frac{1}{2}(\sin^4 x + \cos^2 x) + \frac{1}{2}(\sin^4 y + \cos^2 y) + \frac{1}{2}(\sin^4 z + \cos^2 z)$$

$$\leq \frac{1}{2}(\sin^2 x + \cos^2 x) + \frac{1}{2}(\sin^2 y + \cos^2 y) + \frac{1}{2}(\sin^2 z + \cos^2 z) = \frac{3}{2}.$$

Below we prove that the equality does not hold. Suppose that the equalities holds for some real x, y, z. Then

$$\sin^4 x = \cos^2 y, \quad \sin^4 y = \cos^2 z, \quad \sin^4 z = \cos^2 x,$$

and

$$\sin^2 x, \ \sin^2 y, \ \sin^2 z = 0 \text{ or } 1.$$

From

$$\sin^2 x = 0 \Rightarrow \cos^2 x = 1 \Rightarrow \sin^4 z = 1 \Rightarrow \sin^2 z = 1 \Rightarrow \cos^2 z = 0$$
$$\Rightarrow \sin^4 y = 0 \Rightarrow \sin^2 y = 0 \Rightarrow \cos^2 y = 1 \Rightarrow \sin^4 x = 1 \Rightarrow \sin^2 x = 1,$$

a contradiction.

Similarly, $\sin^2 x = 1$ yields a contradiction also. Thus, the equality is impossible, the conclusion is proven.

3. Let $x_k = \tan^2 \theta_k$ for $k = 1, 2, \ldots, n$, where $\theta_k \in \left[0, \dfrac{\pi}{2}\right)$, and take $\theta_{n+1} = \theta_1$, then

$$\sqrt{\frac{1}{(x_k + 1)^2} + \frac{x_{k+1}^2}{(x_{k+1} + 1)^2}} = \sqrt{\cos^4 \theta_k + \sin^4 \theta_{k+1}}$$

$$\geq \sqrt{\frac{1}{2}(\cos^2 \theta_k + \sin^2 \theta_{k+1})^2} = \frac{\cos^2 \theta_k + \sin^2 \theta_{k+1}}{\sqrt{2}},$$

Thus,

$$\sum_{k=1}^{n} \sqrt{\frac{1}{(x_k + 1)^2} + \frac{x_{k+1}^2}{(x_{k+1} + 1)^2}} \geq \sum_{k=1}^{n} \frac{\cos^2 \theta_k + \sin^2 \theta_{k+1}}{\sqrt{2}} = \frac{n}{\sqrt{2}}.$$

4. The AM-GM inequality gives

$$\frac{\cos^2 a}{\sin^2 a \sin^2 b \cos^2 b} + \frac{\sin^2 a}{\cos^2 a} \geq \frac{2}{\sin b \cos b}$$

for any $0 < a, b < \dfrac{\pi}{2}$, therefore

$$\text{L.H.S.} = \left(\frac{5}{\cos^2 a} + \frac{5}{\sin^2 a \sin^2 b \cos^2 b}\right) \cdot (\cos^2 a + \sin^2 a)$$

$$= 5 + 5\left(\frac{\cos^2 a}{\sin^2 a \sin^2 b \cos^2 b} + \frac{\sin^2 a}{\cos^2 a}\right) + \frac{5}{\sin^2 b \cos^2 b}$$

$$\geq 5 + \frac{10}{\sin b \cos b} + \frac{5}{\sin^2 b \cos^2 b} = 5\left(1 + \frac{1}{\sin b \cos b}\right)^2$$

$$= 5\left(1 + \frac{2}{\sin 2b}\right)^2 \geq 45 \geq 45\sin(a + \varphi) = 27\cos a + 36\sin a,$$

where $\cos \varphi = \dfrac{4}{5}, \sin \varphi = \dfrac{3}{5}$.

5. Note that

$$\frac{\sin^n a + \sin^n b}{(\sin a + \sin b)^n} \geq \frac{\sin^n 2a + \sin^n 2b}{(\sin 2a + \sin 2b)^n}$$

$$\Leftrightarrow (\sin^n a + \sin^n b)(\sin 2a + \sin 2b)^n \geq (\sin a + \sin b)^n (\sin^n 2a + \sin^n 2b)$$

$\Leftrightarrow (\sin a \sin 2a + \sin a \sin 2b)^n + (\sin b \sin 2a + \sin b \sin 2b)^n$

$\geq (\sin a \sin 2a + \sin b \sin 2a)^n + (\sin 2b \sin a + \sin 2b \sin b)^n.$

Letting $\sin a \sin 2a = u, \sin a \sin 2b = v, \sin b \sin 2a = x, \sin b \sin 2b = y$. It suffices to show that

$$(u + v)^n + (x + y)^n \geq (u + x)^n + (v + y)^n.$$

Without loss of generality, we may assume that $a \geq b$. Then $1 > u \geq v \geq x \geq y > 0$. Thus

$(u + v)^n + (x + y)^n \geq (u + x)^n + (v + y)^n$
$\Leftrightarrow (u + v)^n - (u + x)^n \geq (v + y)^n - (x + y)^n$
$\Leftrightarrow (v - x)[(u + v)^{n-1} + (u + v)^{n-2}(u + x) + \cdots + (u + x)^{n-1}]$
$\quad \geq (v - x)[(v + y)^{n-1} + (v + y)^{n-2}(x + y) + \cdots + (x + y)^{n-1}].$

Since

$$(u + v)^k (u + x)^{n-1-k} \geq (v + y)^k (x + y)^{n-1-k} \quad \text{for } 0 \leq k \leq n - 1,$$

the conclusion is proven at once by adding them up.

Solutions to Testing Questions   10

Testing Question   (10-A)

1. Let M and N be the midpoints of AB and CD respectively. Then O, M, N are collinear, and the line ON is the perpendicular bisector of AB and CD. Let $\angle CON = x$, then $\angle AON = 3x$ and

$$\frac{23}{9} = \frac{AM}{CN} = \frac{r \sin 3x}{r \sin x} = 3 - 4\sin^2 x$$

$$\Rightarrow \sin^2 x = \frac{1}{4}\left(3 - \frac{23}{9}\right) = \frac{1}{9}$$

$$\Rightarrow \sin x = \frac{1}{3},$$

$$\therefore r = \frac{CN}{\sin x} = 27.$$

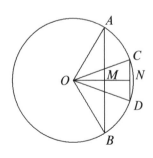

2. There are two possible cases: (i) As shown the following left figure, when the given point A is outside the given circle O, $AB = 4$ cm, $AC = 9$ cm, so $BC = 5$ cm, i.e. the radius r is 2.5 cm.

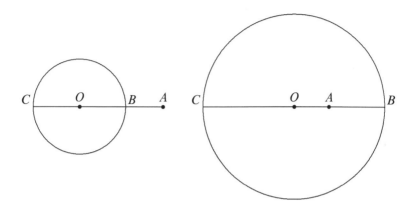

(ii) When the given point A is inside the given circle O, as shown in the right figure above, then $AB = 4$ cm, $AC = 9$ cm, so $BC = 13$ cm, i.e., $r = 6.5$ cm.

Therefore r is 2.5 cm or 6.5 cm, the answer is (B).

3. Since $POAB$ is a parallelogram, $AB \parallel PO$, and $AB = PO$, which implies that

$SOBA$ is also a parallelogram, so

$$BO \parallel QS.$$

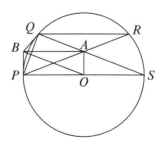

Since PS is a diameter of the circle, so $PQ \perp QS$, i.e., $BO \perp PQ$.
By the midpoint Theorem, BO passes through the midpoint of PQ, so it is the perpendicular bisector of PQ.
Thus, $BQ = BP$.

4. As shown in the figure below, $\angle BAP = \angle BCP$ and $\angle ABC = \angle APC$

implies $\triangle ABD \sim \triangle CPD$ (A.A.A.),

$$\therefore \frac{PD}{BD} = \frac{CP}{AB}.$$

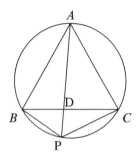

Since

$$AB = AC \Rightarrow \angle BPA = \angle CPA,$$

By the angle bisector theorem,

$$\frac{BD}{CD} = \frac{BP}{CP} = \frac{21}{28} = \frac{3}{4},$$

therefore $PD = BD \cdot \dfrac{CP}{AB} = CP \cdot \dfrac{BD}{AB} = CP \cdot \dfrac{BD}{BC} = 28 \cdot \dfrac{3}{7} = 12.$

5. As shown in the given figure below, the given condition implies

$$\overset{\frown}{APB} = \angle APB = \frac{1}{2}\, \overset{\frown}{ADB},$$

where D and P are on different sides of AB, therefore

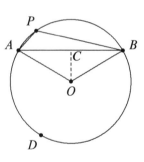

$$\overset{\frown}{APB} = \tfrac{1}{3} \cdot 360° = 120°,$$
$$\angle AOB = \overset{\frown}{APB} = 120°.$$

Let $OC \perp AB$ at C, then $AC = CB$, so

$$AB = 2AC = 2\sin 60° = \sqrt{3}.$$

6. We are given that $AB < AD$. Since CY bisects $\angle BCD$, $BY = YD$, so Y

lies between D and A on the circle, as in the right diagram, and $DY > YA, DY > AB$. Similar reasoning confirms that X lies between B and C and $BX > XC, BX > CD$. So if $ABXCDY$ has 4 equal sides, then it must be that $YA = AB = XC = CD$. This implies that $\overset{\frown}{YAB} = \overset{\frown}{XCD}$ and hence that $YB = XD$. Since $\angle BAX = \angle XAD$,

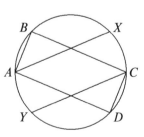

so $BX = XD$. Since $\angle DCY = \angle YCB$, so $DY = YB$. Therefore $BXDY$ is a square and its diagonal, BD, must be a diameter of the circle.

7. Let R be the radius of the circumcircle, r be the radius of the inscribed circle

and s be the perimeter of the convex n-sided polygon.
Since the convex n-sided polygon is inscribed in its circumcircle, so

$$s < 2\pi R.$$

therefore

$$B = \frac{1}{2}sr < \pi Rr.$$

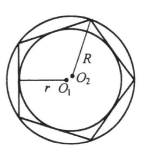

Thus,

$$
\begin{aligned}
2B \quad &< \quad 2\pi Rr \\
&= \quad \pi \cdot 2Rr < \pi(R^2 + r^2) \\
&= \quad \pi R^2 + \pi r^2 = A + C.
\end{aligned}
$$

8. As shown in the diagram below, Let D, E be the midpoints of AB and AC respectively, then $FD \perp AB$ at D and $GE \perp AC$ at E. Therefore

$$BF = \frac{BD}{\sin \angle BFD} = \frac{BD}{\sin \angle ABC} = \frac{AB}{2\sin \angle ABC}.$$

Similarly, $CG = \dfrac{AC}{2\sin \angle ACB}$. Thus,

$$
\begin{aligned}
BF \cdot CG &= \frac{AB \cdot AC}{4\sin \angle ABC \cdot \sin \angle ACB} \\
&= \frac{4R^2 \sin \angle ACB \cdot \sin \angle ABC}{4\sin \angle ABC \cdot \sin \angle ACB} = R^2.
\end{aligned}
$$

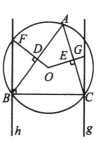

9. (i) $\angle ABC = 120°$ and $\angle ABD = 30°$ gives $\angle DBC = 90°$, therefore

DC passes through the center O. Connect
OA. Then $\angle DCA = \angle ABD = 30°$ and
$OA = OC$ implies that

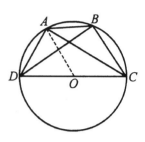

$$AD = AO = DO = \frac{c}{2}.$$

Let $\angle BDC = \theta$, then $\angle ADB = 60° - \theta$.
Applying the sine rule to $\triangle ABD$ gives

$$a = \frac{c}{2} \cdot \frac{\sin(60° - \theta)}{\sin 30°} = c\sin(60° - \theta).$$

In the $\triangle BCD$, $b = c\sin\theta$ yields that

$$\begin{aligned} a + b &= c\sin(60° - \theta) + c\sin\theta = 2c\sin 30° \cos(30° - \theta) \\ &= c\cos(30° - \theta) \le c. \end{aligned}$$

(ii) $|\sqrt{c+a} - \sqrt{c+b}| = \sqrt{c-a-b} \Leftrightarrow 2c + a + b - 2\sqrt{(c+a)(c+b)} = c - a - b \Leftrightarrow 2a + 2b + c = 2\sqrt{(c+a)(c+b)} \Leftrightarrow a^2 + b^2 + ab = \frac{3}{4}c^2$.

Applying the cosine rule to $\triangle ABC$ gives $AC^2 = a^2 + b^2 + ab$, and
applying the Pythagoras' Theorem to $\triangle ADC$ gives $AC^2 = c^2 - \left(\frac{c}{2}\right)^2 = \frac{3}{4}c^2$, hence $a^2 + b^2 + ab = \frac{3}{4}c^2$.

10. Considering PD as the altitude of the $\text{Rt}\triangle PBC$ on the hypotenuse BC,
then the projection theorem gives

$$PC^2 = CD \cdot CB.$$

Similarly, $QC^2 = CE \cdot CA$. Therefore it
suffices to show

$$CD \cdot CB = CE \cdot CA.$$

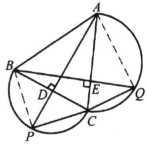

Since $\text{Rt}\triangle BCE \sim \text{Rt}\triangle ACD$ (A, A),
therefore

$$\frac{CD}{CE} = \frac{CA}{CB}$$

gives $CD \cdot CB = CE \cdot CA$ at once. Thus, the conclusion is proven.

Testing Questions   (10-B)

1. Since

    $$\angle ADP = \angle ACB = \angle APB \text{ and}$$
    $$\angle DAP = \angle PAB,$$
    $$\therefore \triangle ADP \sim \triangle APB.$$

    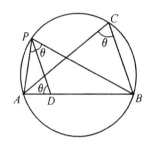

    Since $[APB] : [ADP] = AD : AB = 4 : 1$, so

    $$\frac{PB}{PD} = \sqrt{4} = 2.$$

2. (i)   Since $EH + HF = k + 2$ and $EH - HF = 2$, $EH = \frac{1}{2}k + 2, HF = \frac{1}{2}k$. Then

    $$EH \cdot HF = 4k \Rightarrow k = 12, EH = 8, HF =$$

    (ii)   Connect BD. Then $\angle ADB = \alpha$, so

    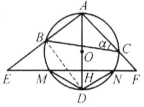

    $$\frac{AB}{BD} = \tan \alpha = \frac{3}{4}.$$

    Write $AB = 3m, BD = 4m$, where $m > 0$, then $\angle ABD = 90° \Rightarrow AD = 5m$. $MD = ND$ implies that $AD \perp EF$ at H, so $\text{Rt} \triangle AHE \sim \text{Rt} \triangle ABD$, therefore

    $$\frac{AB}{AH} = \frac{BD}{EH} \Leftrightarrow \frac{3m}{\frac{3}{4}AD} = \frac{4m}{8} \Rightarrow AD = 8,$$

    so $m = \frac{8}{5}, AH = 6, AE = \sqrt{8^2 + 6^2} = 10, AF = \sqrt{6^2 + 6^2} = 6\sqrt{2}$ and $AB = \frac{24}{5}$. In $\triangle ABC$ and $\triangle AEF$,

    $$\angle ACB = \angle ADB = \angle AEF, \angle BAC = \angle FAE \Rightarrow \triangle ABC \sim \triangle AEF,$$

    therefore $BC = EF \cdot \frac{AB}{AF} = (8 + 6) \cdot \frac{24}{5 \cdot 6\sqrt{2}} = \frac{28\sqrt{2}}{5}$.

3. (i)   For any right triangle of sides a, b, c with $c > a, b$, the length of diameter of its inscribed circle is $d = a + b - c$. Since $c = \sqrt{a^2 + b^2} \geq \frac{a + b}{\sqrt{2}}$, so $d \leq a + b - \frac{a + b}{\sqrt{2}} = \frac{2 - \sqrt{2}}{2}(a + b)$.

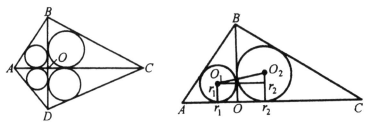

Let d_1, d_2, d_3, d_4 bee the diameters of the $\odot O_1, \odot O_2, \odot O_3, \odot O_4$ respectively, then

$$d_1 \leq \frac{2-\sqrt{2}}{2}(AO + BO), \quad d_2 \leq \frac{2-\sqrt{2}}{2}(BO + CO),$$

$$d_3 \leq \frac{2-\sqrt{2}}{2}(CO + DO), \quad d_4 \leq \frac{2-\sqrt{2}}{2}(DO + AO),$$

by adding them up the conclusion is obtained at once:

$$d_1 + d_2 + d_3 + d_4 \leq (2 - \sqrt{2})(AC + BD).$$

(ii)   Let the radii of the $\odot O_1$ and $\odot O_2$ be r_1 and r_2 respectively. Then

$$O_1 O_2 = \sqrt{(r_1 + r_2)^2 + (r_1 - r_2)^2} = \sqrt{2(r_1^2 + r_2^2)} < \sqrt{2}(r_1 + r_2),$$

and similarly,

$$O_2 O_3 < \sqrt{2}(r_2 + r_3), \quad O_3 O_4 << \sqrt{2}(r_3 + r_4), \quad O_4 O_1 < \sqrt{2}(r_4 + r_1).$$

By adding up these four inequalities and using the result of (i), it is obtained that

$$\begin{aligned} O_1 O_2 + O_2 O_3 + O_3 O_4 + O_4 O_1 \quad &< \quad \sqrt{2}(d_1 + d_2 + d_3 + d_4) \\ &\leq \quad 2(\sqrt{2} - 1)(AC + BD). \end{aligned}$$

4.   As shown in the figure below, let Z be the point of intersection of BC and Γ_1. Connect MZ.

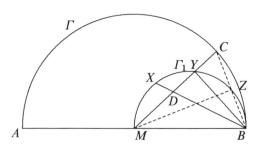

Since $\angle MZB = 90°$ and $MC = MB$ implies that Z is the midpoint of BC and $\angle YMZ = \angle BMZ$, so M is also the midpoint of the arc $Y\overset{\frown}{Z}B$.

Since $BY \perp MC$ and $\overset{\frown}{XY} = \overset{\frown}{YZ} = \overset{\frown}{ZB}$, so $\angle YBX = \angle YBC$ which implies that $\text{Rt}\triangle YBX \cong \text{Rt}\triangle YBC$, hence $CY = YD$.

5. Let M be the midpoint of AC, and F' the second point of intersection of the line BM and the circle Ω. Let B' be the point of intersection of the line FM and Ω.

    $\angle ABE = \angle CBE$ implies that E is the midpoint of the arc AC, so E, M both are on the perpendicular bisector l of the segment AC. Therefore $\angle EMD = 90°$ and M is on the circle ω. Hence

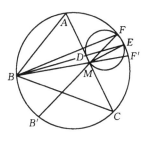

$$\frac{1}{2}\,\overset{\frown}{B'CE} = \angle B'FE = \angle MFE$$
$$= \angle MDE = \angle CDE = \frac{1}{2}(\overset{\frown}{AB} + \overset{\frown}{CE})$$
$$= \frac{1}{2}(\overset{\frown}{AB} + \overset{\frown}{AE}) = \frac{1}{2}\,\overset{\frown}{BAE}\,.$$

Therefore B and B' are symmetric with respect to l. It implies that F and F' are symmetric with respect to l, $\overset{\frown}{FE} = \overset{\frown}{F'E}$, $\angle FBE = \angle F'BE$, so BF and BM are symmetric with respect to BE.

Solutions to Testing Questions   11

Testing Questions   (11-A)

1. Without loss of generality, we assume that $AB = 1$. Connect BD. From E introduce $EF \perp BD$ at F. $\overset{\frown}{AP} = \overset{\frown}{PB}$ implies that

    $$\angle ADP = \angle PDB \Rightarrow \text{Rt}\triangle EAD \cong \text{Rt}\triangle EFD,$$

    $$\therefore EF = EA. \quad \text{Consider the Rt}\triangle BEF,$$

    $$\angle EBF = 45° \Rightarrow BE = \sqrt{2}EF = \sqrt{2}AE.$$

Let $AE = x$, then $BE = 1 - x = \sqrt{2}x$, so $x = \dfrac{1}{\sqrt{2}+1} = \sqrt{2} - 1$, then

$BE = 2 - \sqrt{2}$. By the Pythagoras' theorem,

$$DE = \sqrt{AE^2 + AD^2} = \sqrt{4 - 2\sqrt{2}},$$

then the intersecting chords theorem yields

$$PE = \frac{AE \cdot BE}{DE} = \frac{(\sqrt{2}-1)(2-\sqrt{2})}{\sqrt{4-2\sqrt{2}}} = \frac{3\sqrt{2}-4}{\sqrt{4-2\sqrt{2}}}.$$

Thus,

$$\frac{PE}{DE} = \frac{3\sqrt{2}-4}{\sqrt{4-2\sqrt{2}}} \div \sqrt{4-2\sqrt{2}} = \frac{3\sqrt{2}-4}{4-2\sqrt{2}} = \frac{\sqrt{2}-1}{2}.$$

2. Since the diameter AD is perpendicular to the chord BC so it bisects BC, i.e., $BE = EC = \sqrt{5}$ in cm.

    $$BD \parallel CF \Rightarrow \angle DBE = \angle FCE \Rightarrow \text{Rt}\triangle DBE \cong \text{Rt}\triangle FCE,$$

    so that $DE = FE$. Thus, $OF = FE = DE$, and, letting $ED = x$, $AE = 5x$. By the intersecting chords theorem,

    $$AE \cdot ED = BE \cdot EC \Rightarrow 5x^2 = 5 \Rightarrow x = 1.$$

    Thus, $CD = \sqrt{BE^2 + ED^2} = \sqrt{5+1} = \sqrt{6}$.

3. Let $R > r$ and they are the radii of the circles $\odot P$ and $\odot Q$ respectively. Let C is the tangent point of the circles. One external common tangent is tangent to the two circles at A and B respectively, and the segment on the internal common tangent which is between the two external common tangents is DE.

    Suppose that $QS \parallel BA$, intersecting AP at S, then $SQBA$ is a rectangle. Since $CD = AD = BD$ and $CD = DE$ by symmetry, so

    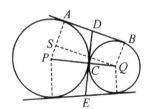

    $$\begin{aligned} DE &= 2DC = AB = \sqrt{PQ^2 - PS^2} \\ &= \sqrt{(R+r)^2 - (R-r)^2} \\ &= \sqrt{4Rr} = 2\sqrt{Rr}. \end{aligned}$$

4. Let $ABCD$ be the quadrilateral, as shown in the given diagram. It has an inscribed circle implies that $AB + CD = AD + BC$. Since $AK_1 = AN_2, BK_2 = BL_1, CL_2 = CM_1, DM_2 = DN_1$, therefore

$$K_1K_2 + M_1M_2 = L_1L_2 + N_1N_2.$$

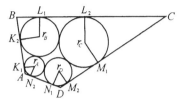

Let r_A, r_B, r_C, r_D be the radii of the four circles respectively, then (cf. the result of Q3. above)

$$\sqrt{r_A r_B} + \sqrt{r_C r_D} = \sqrt{r_B r_C} + \sqrt{r_A r_D},$$

so $(\sqrt{r_A} - \sqrt{r_C})(\sqrt{r_B} - \sqrt{r_D}) = 0$. Thus, $r_A = r_C$ or $r_B = r_D$.

5. Let r and r_1 be the radii of ω and ω_1 respectively. Without loss of generality, we may assume that the tangent point of the two circles is on the arc AB and closer to A.

Let O and O_1 be the centers of ω and ω_1 respectively. Write $\angle O_1 OA = \alpha$, then

$$\angle O_1 OB = 120° - \alpha, \angle O_1 OC = 120° + \alpha.$$

The cosine rule gives

$$
\begin{aligned}
AA_1^2 &= AO_1^2 - r_1^2 = r^2 + (r + r_1)^2 - 2r(r + r_1)\cos\alpha - r_1^2 \\
&= 2r(r + r_1)(1 - \cos\alpha) = 4r(r + r_1)\sin^2\tfrac{\alpha}{2}. \\
\therefore AA_1 &= 2\sin\frac{\alpha}{2} \cdot \sqrt{r(r + r_1)}.
\end{aligned}
$$

Similarly,

$$
\begin{aligned}
BB_1 &= 2\sin\left(60° - \tfrac{\alpha}{2}\right)\sqrt{r(r + r_1)}, \\
CC_1 &= 2\sin\left(60° + \tfrac{\alpha}{2}\right)\sqrt{r(r + r_1)}, \\
\therefore AA_1 + BB_1 &= 2\sqrt{r(r + r_1)}\left[\sin\frac{\alpha}{2} + \sin\left(60° - \frac{\alpha}{2}\right)\right] \\
&= 2\sqrt{r(r + r_1)}\sin\left(60° + \frac{\alpha}{2}\right) = CC_1.
\end{aligned}
$$

6. As shown in the graph below, let O be the center of the big circle and A, B be centers of two adjacent small circles which are tangent externally and both are tangent to the $\odot O$ internally. In $\triangle OAB$, $AB = 2, AO = BO = 10$, therefore, letting $\dfrac{\angle AOB}{2} = \theta$, then $\sin\theta = \dfrac{1}{10}$ and $\tan\theta = \dfrac{1}{\sqrt{99}}$.

n small circles can be put in if and only if

$$2n\theta \le 2\pi \qquad \text{or} \qquad n \le \frac{\pi}{\theta}.$$

Since $\sin\theta < \theta < \tan\theta$, if $2n\tan\theta \le 2\pi$ then

$$\frac{\pi}{\tan\theta} = \sqrt{99}\pi \approx 31.1 \Rightarrow n \ge 31,$$

so 31 small circles can be put in.

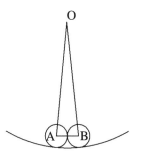

Since $2n\sin\theta \ge 2\pi \Rightarrow 2n\theta > 2\pi$, and $\dfrac{\pi}{\sin\theta} = 10\pi \approx 31.4$ so $n < 32$.
Thus, $31 \le n < 32$, i.e., $n = 31$, the answer is (B).

7. Suppose that CE is tangent to $\odot M$ at N. Connect CM, EM and MN.
    $\angle MDC = \angle MNC = 90°$ and $MD = MN$ implies that

    $$\triangle MNC \cong \triangle MDC.$$

    Similarly, $\triangle MNE \cong \triangle MAE$. Therefore

    $$\angle CMN = \angle CMD, \angle EMN = \angle EMA,$$

    so that $\angle EMC = 90°$, $CN = CD = 1$. By the projection theorem,

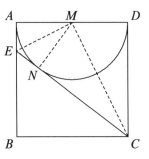

$$MN^2 = EN \cdot CN \Rightarrow EN = \frac{1}{4} \Rightarrow AE = EN = \frac{1}{4}, BE = \frac{3}{4},$$

$$\text{hence } [BEC] = \frac{1}{2} \cdot BE \cdot BC = \frac{3}{8}.$$

8. The digram below shows the upper half of the rhombus. The given conditions
    implies that $\triangle ABC$ is isosceles with $AB = BC$. Let X, Y, Z be the tan-
    gent points of the inscribed circle to AB, BC and EF respectively. Write

$$\angle XOE = \angle EOZ = \alpha,$$
$$\angle ZOF = \angle FOY = \beta,$$
$$\angle AOX = \angle COY = \gamma.$$

Then $\alpha + \beta + \gamma = 90°$, $\angle AEO = 90° - \alpha = \beta + \gamma = \angle COF$, therefore

$$\triangle AOE \sim \triangle CFO,$$

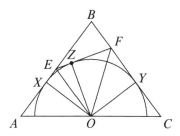

which gives $AE \cdot CF = AO \cdot OC = AO^2$. Similarly, on the lower half of the rhombus we have $AH \cdot CG = AO^2$, therefore

$$\frac{AE}{AH} = \frac{CG}{CF}$$

which implies that $\triangle AEH \sim \triangle CGF$. Then $AB \parallel CD$ implies $EH \parallel FG$.

9. Let AC, EG intersect at a point O. Take the point K on EB such that

$\angle EOK = \angle GOC$. Then

$$\angle OEK = \angle OGC \Rightarrow \triangle OEK \sim \triangle OGC$$

$$\Rightarrow \frac{OK}{KE} = \frac{OC}{CG}.$$

$$\because \angle AOE = \angle COG = \angle KOE,$$

$$\therefore \frac{OA}{AE} = \frac{OK}{KE} = \frac{OC}{CG} \Rightarrow \frac{AO}{OC} = \frac{AE}{CG}.$$

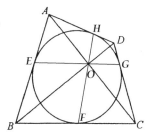

Let AC and HF intersect at O', then similarly $\dfrac{AO'}{O'C} = \dfrac{AH}{CF}$. Since $AE = AH$ and $CG = EG$, so $\dfrac{AO}{OC} = \dfrac{AO'}{O'C}$, i.e. $O = O'$. Thus, AC, EG, HF are concurrent at O.

Similarly, BD, EG, HF are concurrent at one common point, and it is the point of intersection of EG and HF, namely the point O. The conclusion is proven.

Testing Questions   (11-B)

1. As shown in the right diagram below, connect AO, AD, DO and DQ. Then $AP = AQ$ since AP and AQ are two tangents from P to $\odot O$. Since OP, OQ are both radii of the $\odot O$,

    $$\angle APO = \angle AQO = 90°.$$

    By symmetry, $\angle POQ = 2\angle AOQ$ and $OA \perp PQ$ at M. Applying the projection theorem of right triangles gives

    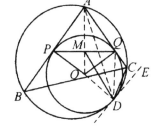

    $$OD^2 = OQ^2 = OM \cdot OA,$$

    i.e., $\dfrac{OD}{OM} = \dfrac{OA}{OD}$. Besides, $\angle AOD = \angle DOM$ is shared, so $\triangle DOM \sim \triangle AOD$, hence $\angle ODM = \angle OAD$. By passing through D introduce the common tangent line DE, then $\angle CDE = \angle CAD$. Since $OD \perp DE$, so

    $$\begin{aligned} \angle MDC &= 90° - \angle ODM - \angle CDE = 90° - \angle OAD - \angle DAC \\ &= 90° - \angle OAQ = \angle AOQ. \end{aligned}$$

    Thus, $\angle POQ = 2\angle MDC$.

2. Let $MP \perp AC$ at P and $MQ \perp AB$ at Q. Suppose that the $\odot I$ touches BC at D. Then $ID \perp BC$ at $D, IF \perp AB, IE \perp AC$. Since $AF = AE$, so $\angle AFM = \angle AEM$, therefore

    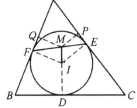

    $$\text{Rt}\triangle QFM \sim \text{Rt}\triangle PEM, \quad \frac{MQ}{MP} = \frac{MF}{ME}.$$

    $$\therefore \ \frac{[MAB]}{[MAC]} = \frac{MQ \cdot AB}{MP \cdot AC} = \frac{MF}{ME} \cdot \frac{AB}{AC},$$

    therefore $[MAB] = [MAC]$ if and only if

    $$\frac{ME}{MF} = \frac{AB}{AC}. \tag{$*$}$$

    Below we show that $(*)$ holds if and only if $MI \perp BC$.

    Suppose that $MI \perp BC$. Then M, I, D are collinear. Since the quadrilaterals $BDIF$ and $CDIE$ are both cyclic,

    $$\angle MIF = \angle B, \qquad \angle MIE = \angle C.$$

Since $IE = IF$, by the sine rule,

$$\frac{MF}{\sin \angle MIF} = \frac{FI}{\sin \angle IMF} = \frac{IE}{\sin \angle IME} = \frac{ME}{\sin \angle MIE},$$

so $\dfrac{ME}{MF} = \dfrac{\sin C}{\sin B} = \dfrac{AB}{AC}$ since the sine rule again.

Conversely, when $\dfrac{AB}{AC} = \dfrac{ME}{MF}$, let M' be the point of intersection of the line DI and EF. Then above proof indicates that $\dfrac{AB}{AC} = \dfrac{M'E}{M'F}$, therefore $\dfrac{M'E}{M'F} = \dfrac{ME}{MF}$, and which implies $\dfrac{M'E}{EF} = \dfrac{ME}{EF}$, so $M'E = ME$, namely M' coincides with M, so $MI \perp BC$.

3. Let $AM = x, CM = 15 - x, BM = d$, then $\dfrac{[ABM]}{[CBM]} = \dfrac{x}{15 - x}$. Let $BM = d$ and p_1, p_2 be the perimeters of $\triangle ABM$ and $\triangle CBM$ respectively, then

$\dfrac{x}{15 - x} = \dfrac{p_1}{p_2} = \dfrac{12 + d + x}{28 + d - x}$

$\Rightarrow 25x + 2dx = 15d + 180$, therefore

$d = \dfrac{25x - 180}{15 - 2x}$. Since $d > 0$, so

$\dfrac{36}{5} < x < \dfrac{15}{2}$.

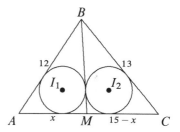

On the other hand, by Stewart's Theorem, $d^2 = \dfrac{12^2(15 - x) + 13^2 x}{15} - x(15 - x)$, so $15d^2 + 15x(15 - x) = 144(15 - x) + 169x$ which gives $432 = 3d^2 + 40x - 3x^2$, therefore $432 = \dfrac{3(25x - 180)^2}{(15 - 2x)^2} + 40x - 3x^2$, or $12x^4 - 340x^3 + 2028x^2 - 7920x = 0$. By factorization,

$$4x(x - 15)(3x^2 - 40x + 132) = 0,$$
$$4x(x - 15)(x - 6)(3x - 22) = 0,$$

$$\therefore x = \frac{22}{3}.$$

Thus, $AM = \dfrac{22}{3}, CM = 15 - \dfrac{22}{3} = \dfrac{23}{3}$ and $\dfrac{AM}{CM} = \dfrac{22}{23}$, so $p + q = 45$.

4. Let K be the second point of intersection of the line ADE with Γ_1. Then well known by us that $\triangle KFD \sim \triangle BAD$.

Since $\angle ABD = \angle DEB = \angle AEB$,

$$\therefore \triangle BAD \sim \triangle EAB,$$

then $\triangle KFD \sim \triangle EAB$. Thus,

$$\frac{FD}{DK} = \frac{AD}{DB} = \frac{AB}{BE}.$$

Let O be the center of Γ_2. $OB \perp AB$ implies that AO is the diameter of Γ_1, so $AK \perp AK$, and OK is the height line of the isosceles triangle ODE, so $DK = KE$. Thus,

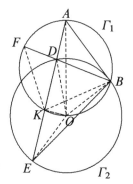

$$EB \text{ is tangent to } \Gamma_1 \text{ at } B \Leftrightarrow \angle EBK = \angle BAK = \angle BAD$$

$$\Leftrightarrow \triangle EKB \sim \triangle BDA \Leftrightarrow \frac{AB}{BE} = \frac{DB}{KE} \Leftrightarrow \frac{FD}{DK} = \frac{DB}{KE} \Leftrightarrow FD = DB.$$

Solutions to Testing Questions   12

Testing Questions   (12-A)

1. Since the given conditions give $\angle AMC = \angle BAC = \angle BPC$, therefore the quadrilateral $BMCP$ is cyclic, hence

$$\begin{aligned} \angle MPA &= \angle MPC - \angle APC \\ &= \angle MBC - \angle CAP \\ &= \angle ABC + \angle ACB - 90°. \end{aligned}$$

Similarly, $\angle NPA = \angle ACB + \angle ABC - 90°$.
Thus, PA bisects the $\angle MPN$.

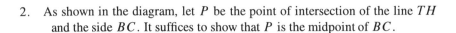

2. As shown in the diagram, let P be the point of intersection of the line TH and the side BC. It suffices to show that P is the midpoint of BC.

Since $\angle AHC = \angle ATC = 90°$, $AHTC$ is cyclic, therefore

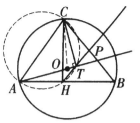

$$\angle PHC = \angle TAC = \frac{1}{2}(180° - \angle AOC)$$

$$= 90° - \angle ABC = \angle PCH,$$

so $\triangle PHB$ is isosceles with $PC = PH$. Since BHC is a right triangle, it gives

$PC = PH = PB$, i.e., P is the midpoint of BC.

3. Let O be the center of the circle Γ. Since M bisects $\overparen{BAC}$, if the line AK intersects Γ again at N, then MN is the diameter of Γ. Therefore

$$MA \perp NA, MA' \perp NA',$$

hence K is the orthocenter of $\triangle MRN$, and A, A', R, K are all on thee circle taking RK as the diameter.

Note that, by the alternate segment theorem,

$$\angle ATA' = 180° - \angle TAA' - \angle TA'A = 180° - \angle AMA' - \angle ANA'$$
$$= 2\angle ARA'$$

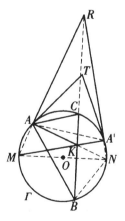

and $TA = TA'$, so T is the center of the circle passing through A', R, A, K.

Thus, T is on RK, namely T, R, K are collinear.

4. First of all we show that A, B, C, D are concyclic. There are two possible cases, as shown in the left diagram below.

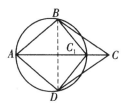 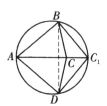 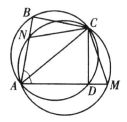

Otherwise, suppose that the circumcircle of $\triangle ABD$ intersects AC at C_1 with $C \neq C_1$. Then $\angle BAC_1 = \angle BAC = \angle DAC = \angle DAC_1$ implies that $\overparen{BC_1} = \overparen{DC_1} \Rightarrow BC_1 = DC_1$. So $BC = DC$ implies that CC_1 is the perpendicular bisector of BD. Then A is on the line CC_1 gives $AB = AD$, a contradiction. Thus, A, B, C, D are concyclic.

As shown in the right diagram above, since the quadrilaterals $ABCD$ and $ANCM$ are both cyclic, $\angle BCD = 180° - \angle DAB = \angle NCM$, therefore

$$\angle NCB = \angle BCM - \angle NCM = \angle BCM - \angle BCD = \angle DCM.$$

Since AC is the angle bisector of $\angle BAD$, so $BC = CD, NC = CM$, hence $\triangle CBN \cong \triangle CDM$. Thus, $DM = BN = a$.

5. Let the circumcircle of $\triangle PEC$ and $\odot O$ intersect at the second point B'. Connect OB', DB', EB', PB', CB'.
    Since P, E, B', C are concyclic and PC is tangent to $\odot O$,

    $$\angle OEB' = \angle B'CP = \angle CDB' = \angle ODB',$$

    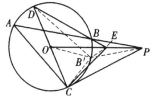

    so O, D, E, B' are concyclic.   Then $OD = OB'$ implies that

    $$\angle DEO = \angle DB'O = \angle ODB' = \angle B'EO.$$

    Therefore if consider OP as a symmetric axis of $\odot O$, then B and B' are symmetric in OP, so that $\angle EBP = \angle EB'P = \angle ECP$. Hence

    $$\angle ACD = \angle ABD = \angle EBP = \angle ECP.$$

    Since $PC \perp CD$, so $AC \perp CE$.

6. Connect OC, BC, then $\angle BOC = \angle BHC = 90°$, so $OHCB$ is cyclic.

Therefore

$$\angle OHB = \angle OCB = 45°.$$

Since $\angle BCM = 90°$, $CH \perp BM$ and M is the midpoint of AC, by the projection theorem of right triangles,

$$AM^2 = CM^2 = MH \cdot MB,$$

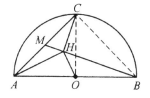

namely, $\dfrac{MH}{MA} = \dfrac{MA}{MB}$, therefore $\triangle AMH \sim \triangle BMA$. Then

$$\angle MAH = \angle MBA, \qquad \angle AHM = \angle BAM = 45°,$$

so that $\angle AHM = \angle BHO$, therefore $\triangle AMH \sim \triangle BOH$, hence

$$\frac{AH}{BH} = \frac{MH}{OH}, \quad \text{i.e. } AH \cdot OH = MH \cdot BH.$$

Thus, $CH^2 = MH \cdot BH$ implies that $CH^2 = AH \cdot OH$, as desired.

7. Suppose that I is the point of intersection of DE and the angle bisector of $\angle C$. Connect AI, CI, CE. Since AE is tangent to the circumcircle of $\triangle ABC$ at A,

$$\angle ACB = 180° - \angle DAE.$$

Then $AD = AE$ implies that

$$
\begin{aligned}
180° - \angle DAE &= \angle ADE + \angle AED \\
&= 2\angle AED = \angle ACI,
\end{aligned}
$$

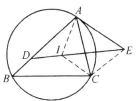

so $\angle ACI = \angle AEI$, $AECI$ is cyclic. Then

$$
\begin{aligned}
\angle IAC &= \angle IEC = \angle AEC - \angle AED \\
&= \frac{180° - \angle CAE}{2} - \frac{180° - \angle DAE}{2} = \frac{\angle DAE - \angle CAE}{2} \\
&= \frac{1}{2}\angle BAC,
\end{aligned}
$$

so AI is the angle bisector of $\angle BAC$, and I is the incenter of $\triangle ABC$, which shows that DE passes through the incenter of $\triangle ABC$.

Testing Questions   (12-B)

1. For a point P on the minor arc $\overset{\frown}{A'B'}$, $\angle APB' + \angle BPA' < 180°$, so the points K, L are both on the major arc $\overset{\frown}{A'C'B'}$. In this case then
    $\angle AKB = 180° - \angle A'KB'$ and
    $\angle ALB = 180° - \angle A'LB'$, so
    $\angle AKB = \angle ALB$, i.e., $AKLB$ is
    cyclic. Let I be the incenter, then

    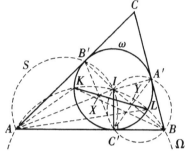

    $$\begin{aligned} \angle AKB &= 180° - \angle A'C'B' \\ &= 180° - \tfrac{1}{2}\angle A - \tfrac{1}{2}\angle B \\ &= \angle AIB, \end{aligned}$$

    thus, A, K, I, L, B are concyclic.

    Let ω_1 and ω_2 be the circumcircles of $AB'IC'$ and $AKILB$ respectively. Then $B'C'$, KL and AI are the common chords of ω and ω_1, ω and ω_2, and ω_1 and ω_2 respectively. They must intersect at one common point X. Further, $X = AI \cap B'C'$ implies that X is the midpoint of $B'C'$, so B' and C' have equal distance from KL. Similarly, KL passes the midpoint Y of $A'C'$, so A' and C' have also equal distance from the line KL, the conclusion now is proven.

2. $AD^2 = AC^2 - CD^2$ let the problem become to show $\dfrac{AE}{BE} = \dfrac{AC \cdot CD}{AD^2}$.
    Let T be the point of intersection of AC and BD. Then

    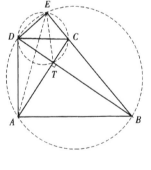

    $$\angle TAD = \angle ABT \Rightarrow \triangle ACD \sim \triangle BDA$$

    $$\Rightarrow \frac{AD}{CD} = \frac{BA}{AD}$$

    $$\Rightarrow AD^2 = AB \cdot CD.$$

    $\angle BAD = \angle BED = 90°$ implies that $ABED$ is cyclic, $\therefore \angle EAD = \angle EBD$. Similarly, $CEDT$ is cyclic, so $\angle CDE = \angle CTE$ which implies that $\angle ADE = \angle BTE$, so $\triangle AED \sim \triangle BET$.

    Therefore $\dfrac{AE}{BE} = \dfrac{AD}{BT}$. But $\triangle ADC \sim \triangle BTA$ implies that $\dfrac{AD}{BT} = \dfrac{AC}{AB}$, therefore

    $$\frac{AE}{BE} = \frac{AC}{AB} = \frac{AC \cdot CD}{AB \cdot CD} = \frac{AC \cdot CD}{AD^2}.$$

3. As shown in the given figure, AC_1 is the common chord of $\odot O$ and $\odot O_2$, so $OO_2 \perp AC_1$.

If $AO_1 \perp OO_2$, then O_1 is on AC_1, therefore

$$OA = OC_1 \Rightarrow \angle OC_1O_1 = \angle OAO_1.$$

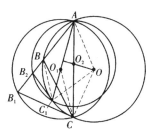

On the other hand, $OA = OC$ and $O_1A = O_1C$ implies that $\angle OCO_1 = \angle OAO_1$,

$$\therefore \angle OC_1O_1 = \angle OCO_1.$$

therefore O, O_1, C_1, C are concyclic. Note that

$$\begin{aligned}
\angle C_1AB_2 &= \frac{1}{2}(\pi - \angle AO_1B_1) = \frac{\pi}{2} - \angle ACB_1 \\
&= \frac{\pi}{2} - \angle BCA - \angle B_1CB = \angle ABC - \frac{\pi}{2},
\end{aligned}$$

then

$$\angle B_2C_1A = \angle B_2C_1B + \angle BC_1A = \angle C_1AB_2 + \angle BCA = \frac{\pi}{2} - \angle CAB.$$

so $\angle AB_2C_1 = \pi - \angle C_1AB_2 - \angle B_2C_1A = \pi + \angle CAB - \angle ABC$. Besides, $\angle B_1CB + \angle BCA = \angle B_1CO_1 + \angle O_1CA$, so

$$\angle CAB + \angle BCA = \left(\frac{\pi}{2} - \angle CAB\right) + \angle C_1AC.$$

Therefore $\angle ABC - \angle CAB = \pi - \left[\left(\frac{\pi}{2} - \angle CAB\right) + \angle C_1AC\right] - \angle CAB = \frac{\pi}{2} - \angle C_1AC < \frac{\pi}{2}$, i.e., $\angle AB_2C_1$ is obtuse. Thus, $\angle OO_2C_1 = \angle AB_2C_1$. Besides,

$$\begin{aligned}
\angle OO_1C_1 &= \angle C_1O_1C + \angle CO_1O = 2\angle C_1AC + \angle CC_1O \\
&= 2\angle C_1AC + \frac{1}{2}(\pi - \angle COC_1) = \frac{\pi}{2} + \angle C_1AC \\
&= \frac{\pi}{2} + \angle CAB - \angle C_1AB_2 = \pi + \angle CAB - \angle ABC.
\end{aligned}$$

Therefore $\angle OO_1C_1 = \angle OO_2C_1$, i.e., O, O_2, O_1, C_1 are concyclic.

Thus, the five points O, O_2, O_1, C_1, C are concylic.

4. When $AB = AC$, then P and Q are both on the angle bisector of $\angle BAC$, so $\angle PAO = \angle QAO = 0$.

When $AB \neq AC$, as shown in the right graph, let O' be the circumcenter of $\triangle ABC$, the perpendicular bisector of BC intersects $\odot O'$ at P' and M, where P' and A are at the same side of BC, then O', O, Q are collinear.

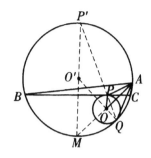

$\angle BAO = \angle CAO \Rightarrow A, O, M$ are collinear. Let R be the second point of intersection of $P'Q$ and $\odot O$, then $\angle MP'R = \angle O'P'Q = \angle O'QR = \angle ORQ$ implies that $MP' \parallel OR$. Then $MP' \perp BC$ implies that $OR \perp BC$, so $P = R$. Thus, P', P, Q are collinear, and $\angle QP'M = \angle PQO = \angle QPO$. Since $\angle QAO = \angle QAM = \angle QP'M$, so $\angle QAO = \angle QPO$, hence A, P, O, Q are concyclic. Thus,

$$\angle PAO = \angle PQO = \angle QPO = \angle QAO.$$

5.   Assume that the disposition of points is as in the diagram.

Since $\angle EBF = 180° - \angle CBF = 180° - \angle EAF$ by hypothesis, the quadrilateral $AEBF$ is cyclic. Hence $AJ \cdot JB = FJ \cdot JE$. In view of this equality, I belongs to the circumcircle of ABK if and only if $IJ \cdot JK = FJ \cdot JE$. Expressing $IJ = IF + FJ$, $JE = FE - FJ$, and $JK = \frac{1}{2}FE - FJ$, we find that I belongs to the circumcircle of ABK if and only if

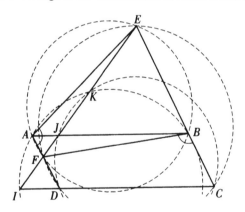

$$FJ = \frac{IF \cdot FE}{2IF + FE}.$$

Since $AEBF$ is cyclic and AB, CD are parallel, $\angle FEC = \angle FAB = 180° - \angle CDF$. Then $CDFE$ is also cyclic, yielding $ID \cdot IC = IF \cdot IE$.

It follows that K belongs to the circumcircle of CDJ if and only if $IJ \cdot IK = IF \cdot IE$. Expressing $IJ = IF + FJ, IK = IF + \frac{1}{2}FE$, and $IE = IF + FE$, we find that K is on the circumcircle of CDJ if and only if

$$FJ = \frac{IF \cdot FE}{2IF + FE}.$$

The conclusion follows.

Solutions to Testing Questions   13

Testing Questions   (13-A)

1. (i)   As shown in the diagram below, let the midpoint of AB be P. The $\odot P$ and $\odot O$ are tangent internally at E implies that P, O, E are collinear. Connect FO, AE, AF, then $OF \perp CD$ at F and $FO \parallel AP$. Since $OE = OF$ and $PE = PA$, so

$$\angle EOF = \angle EPA \Rightarrow \angle OEF = 90° - \frac{\angle EOF}{2} = 90° - \frac{\angle EPA}{2} = \angle PEA,$$

therefore A, F, E are collinear.

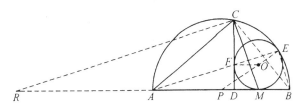

(ii)   By considering the power of A with respect to $\odot O$, we have $AM^2 = AF \cdot AE$. Connect EB. $AE \perp EB$ implies that $EFDB$ is cyclic, so $AF \cdot AE = AD \cdot AB$. Connect BC. By applying the projection theorem to the Rt$\triangle ABC$, then

$$AC^2 = AD \cdot AB = AF \cdot AE = AM^2 \Rightarrow AC = AM.$$

(iii)   Extend MA to R such that $MA = AR$. Connect CR, then $\angle RCM = 90°$, so the projection theorem gives

$$MC^2 = MD \cdot MR = 2MD \cdot MA.$$

2. M is the midpoint of hypotenuse of Rt$\triangle ANP$, so it is the circumcenter of $\triangle ANP$. By the midpoint theorem, $MN \parallel BP$. Let E be the second point of intersection of MN and $\odot O$. By the power of a point theorem,

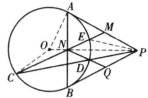

$$PM^2 = MA^2 = ME \cdot MC$$
$$\Rightarrow \frac{PM}{ME} = \frac{MC}{PM}$$
$$\Rightarrow \triangle PME \sim \triangle CMP$$
$$\Rightarrow \triangle MPE \sim \triangle MCP.$$

Since $OAPB$ is cyclic, $CN \cdot NE = AN \cdot NB = ON \cdot NP$, which implies that $CPEO$ is cyclic,

$$\therefore \angle EPN = \angle NCO.$$

Applying the projection theorem to Rt$\triangle OAP$ and considering power of P to $\odot O$, it follows that

$$PD \cdot PC = PA^2 = PN \cdot PO \Rightarrow CDNO \text{ is cyclic} \Rightarrow \angle QNP = \angle PCO$$
$$\Rightarrow \angle QNP = \angle PCM + \angle MCO = \angle MPE + \angle EPN = \angle MPN,$$

hence $MP \parallel NP$, i.e., $MNQP$ is a parallelogram. Then $PQ = PM$ implies that $MNQP$ is a rhombus.

3. Let O be the common circumcenter of the triangles DEF and ABC. Let D', E', F' be the points of intersection of the smaller circle with BC, CA, AB respectively. Make $OH \perp BC$ at H. Let

$$\frac{BD}{DC} = \frac{CE}{EA} = \frac{AF}{FB} = k,$$

then $BH = HC$ and $DH = HD'$ implies that

$$BD' = DC = \frac{1}{k+1}BC.$$

Similarly,

$$BF = \frac{1}{k+1}BA, AF' = \frac{k}{k+1}BA.$$

The power of a point theorem gives $BD' \cdot BD = BF \cdot BF'$, hence we have

$$\frac{k}{(k+1)^2}BC^2 = \frac{k}{(k+1)^2}AB^2, \text{ namely } BC = AB. \text{ Similarly we have}$$
$BC = CA$. Thus, $\triangle ABC$ is equilateral.

4. Let the center and radius of Γ be O and r respectively. Let K be the point of intersection of the circumcircle of $\triangle APS$ and AB, connect SK. Then

$$\angle AKS = \angle SPQ = \angle SRB,$$

so $BKSR$ is cyclic, i.e., the circumcircles of triangles APS and BSR intersect at S and K.

By considering the powers of A and B with respect to the the circles, it follows that

$$AO^2 - r^2 = AS \cdot AR = AK \cdot AB$$
$$= AK^2 + AK \cdot KB$$

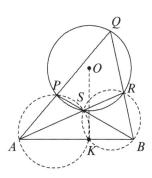

and

$$BO^2 - r^2 = BS \cdot BP = BK \cdot BA = BK^2 + BK \cdot KA.$$

Thus, $AO^2 - AK^2 = BO^2 - BK^2$, which implies that $OK \perp AB$ at K, and $AK \cdot BK = OK^2 - r^2$. Let U, S be two fixed point on the line OK with the distance $\sqrt{OK^2 - r^2}$ from K. Then $AUBS$ is cyclic for each pair $\{A, B\}$ satisfying above conditions, the solution is the set $\{U, S\}$.

5. Let Z_1 be the second point of intersection of Γ_1 with AB, Z_2 be that of Γ_2 with AC, and $\angle CAB = \alpha, \angle ABC = \beta, \angle BCA = \gamma$ in degrees. Then I_1 is the center of Γ_1 implies

$$X_1 Z_1 = PY_1, \quad BZ_1 = BP.$$

Suppose that the inscribed circle of $\triangle ABP$ touches AB, PB at M and N respectively, then M, N are the midpoint of $X_1 Z_1, PY_1$ respectively. Therefore

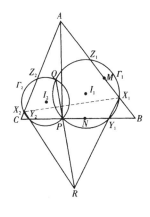

$$AX_1 = AM + \tfrac{1}{2}X_1 Z_1$$
$$= \frac{1}{2}(AP + AB - PB) + \frac{1}{2}X_1 Z_1$$
$$= \frac{1}{2}(AP + AB - PB) + \frac{1}{2}PY_1$$
$$= \frac{1}{2}(AP + AB - PB) + \frac{1}{2}(AP + PB - AB) = AP.$$

Similarly $AX_2 = AP$. Therefore $\triangle AX_1X_2$ is isosceles with $AX_1 = AX_2$. Further, $\triangle BX_1Y_1$ and $\triangle CX_2Y_2$ are also isosceles, so

$$\angle Y_2X_2X_1 = 180° - \left(90° - \frac{\alpha}{2}\right) - \left(90° - \frac{\gamma}{2}\right) = 90° - \frac{\beta}{2}$$

and $\angle Y_2Y_1X_1 = 90° + \dfrac{\beta}{2}$. Hence $X_1Y_1Y_2X_2$ is cyclic.

Suppose that the lines X_1Y_1 and X_2Y_2 intersect at R. Then the power of R with respect to the circumcircle of $X_1Y_1Y_2X_2$ is

$$RX_1 \cdot RY_1 = RX_2 \cdot RY_2,$$

so R is on the radical axis of Γ_1 and Γ_2. Since the axis is just the line PQ, so the lines X_1Y_1, X_2Y_2 and PQ are concurrent at R.

6. Let $B_1 = D$ and the incircle touch the sides BC and AB at A_1 and C_1 respectively.

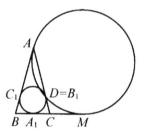

Write M be the tangent point of the second circle to the extension of BC. Let $B_1C = x$, $AB_1 = y$, then

$$A_1M = C_1A = AB_1 = y \quad \text{and} \quad A_1C = B_1C = x$$
$$\Rightarrow CM = y - x.$$

Thus, $CM^2 = CB_1 \cdot CA \Rightarrow (y - x)^2 = x(x + y) \Rightarrow \dfrac{y}{x} = 3.$

Testing Questions   (13-B)

1. Connect AO and extend it to intersect $\odot O$ at A'.

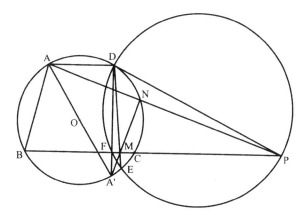

Below we show that M, N, A' are collinear, i.e., the line MN must passes through the fixed point A'.

Let ω be the circle with the diameter PD. Then DE is the radical axis of $\odot O$ and ω. Since $\angle PNA' = 90°$, the line $A'N$ is radical axis of $\odot O$ and the circle with diameter PA' (denoted by ω').

Suppose that DA' intersects BC at F. Since $\angle ADA' = 90°$, so F is on ω and $\angle PFA' = 90°$, then $PNFA'$ is cyclic, i.e., F is on ω' also. Thus, the line BC is radical axis of ω and ω'.

Since the two radical axes DE and BC intersect at M, by the radical center theorem, M is the radical center, namely M is on the line NA', so M, N, A' are collinear.

2. (i) $CL \perp AB$ and $AM \perp AE$ implies that $\angle LEA = \angle BAM$, so

$$AE = \frac{AL}{\sin \angle BAM} = \frac{AC \cos A}{\sin \angle BAM}.$$

Similarly, $AF = \dfrac{AB \cos A}{\sin \angle CAM}.$

Since

$$\frac{AF}{AE} = \frac{AB \sin \angle BAM}{AC \sin \angle CAM} = \frac{BM}{MC} = 1,$$

so $AF = AE$.

(ii)   Since EMF is isosceles, it circumcenter O is on AM. Let O_1 be the circumcenter of Γ_1, and Γ_1 touches Γ and EF at T and D respectively. Then T, O_1, O are collinear. $O_1D \perp EF$ implies that $O_1D \parallel AM$, and $O_1T = O_1D, OT = OM$ implies that $\triangle TO_1D \sim \triangle TOM$, so T, D, M are collinear.

Since $\angle MTE = \angle MFE = \angle MED$, so $\triangle MET \sim \triangle MDE$, therefore $MD \cdot MT = ME^2$. Thus, the power of M to Γ_1 is ME^2. Similarly, the power of M to Γ_2 is ME^2 also, hence M is on the radical axis of Γ_1 and Γ_2, therefore P, Q, M are collinear if Γ_1 and Γ_2 intersect at P, Q.

3. Suppose that the inscribed circle ω of $ABCD$ touches the sides $AB, BC,$

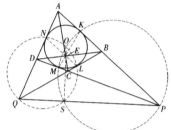

CD, DA at K, L, M, N respectively, as shown in the right diagram. From O make $OS \perp PQ$ at S.
By Newton's Theorem (cf. Q9 in Testing Question 11A), the point of intersection of KM and LN is also E.

On the other hand, $OKPM$ and $OLQN$ are both cyclic, and their circumcircles ω_1 and ω_2 intersect at O and S (cf. Q4 in Testing Question 13A), so the lines OS, KM, NL are the radical axes of ω_1 and ω_2, ω_1 and ω, ω_2 and ω respectively. Hence E is their radical center. Thus,

$$OE \cdot k = OE \cdot OS = OE(OE + ES) = OE^2 + OE \cdot ES$$
$$= OE^2 + LE \cdot EN = OE^2 + r^2 - OE^2 = r^2.$$

4. By S_1, S_2 we denote the circumcircles of $\triangle ACA_1$ and $\triangle BCB_1$ respectively. When the second point of intersection of the two circles, D, is on the side AB, the power of B to S_1 and the power of A to S_2 give

$$AB \cdot BD = BC \cdot BA_1$$

and

$$AB \cdot AD = AC \cdot AB_1.$$

Adding them up yields

$$AB^2 = AB(BD + AD) = BC \cdot BA_1 + AC \cdot AB_1$$
$$= BC(BC - CA_1) + AC(AC - CB_1)$$
$$= BC^2 + AC^2 - (BC \cdot CA_1 + AC \cdot CB_11).$$

Applying the cosine rule to $\triangle ABC$ yields

$$AB^2 = BC^2 + AC^2 - 2BC \cdot AC \cos 60° = BC^2 + AC^2 - BC \cdot AC,$$

so that $BC \cdot CA_1 + AC \cdot CB_1 = BC \cdot AC$, i.e., $\dfrac{CB_1}{CB} + \dfrac{CA_1}{CA} = 1$.

Conversely, when $\dfrac{CB_1}{CB} + \dfrac{CA_1}{CA} = 1$, suppose that AB intersects S_1, S_2 at D_1, D_2 respectively, as shown in the diagram below, then the power theorem gives

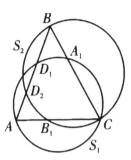

$AB \cdot BD_1 = BC \cdot BA_1$ and $AB \cdot AD_2 = AC \cdot AB_1$. Adding up them gives

$$AB(BD_1 + AD_2) = BC \cdot BA_1 + AC \cdot AB_1$$
$$= BC(BC - CA_1) + AC(AC - CB_1)$$
$$= BC^2 + AC^2 - (BC \cdot CA_1 + AC \cdot CB_1).$$

Then $BC \cdot CA_1 + AC \cdot CB_1 = BC \cdot AC$ yields

$$BC^2 + AC^2 - (BC \cdot CA_1 + AC \cdot CB_1)$$
$$= BC^2 + AC^2 - BC \cdot AC$$
$$= BC^2 + AC^2 - 2BC \cdot AC \cos C = AB^2.$$

Thus, $AB(BD_1 + AD_2) = AB^2$, so $AB = BD_1 + AD_2$ which implies that D_1 coincides with D_2, so the second point of intersection of S_1 and S_2 is on the side AB.

Solutions To Testing Questions   14

Testing Questions   (14-A)

1. Suppose that $AC \cap BD = P, AF \cap BE = M, CE \cap DF = N$. Suppose also that $BD \cap CE = H$.

Since the three points N, P, M are on the lines EH, HB, BE respectively, it suffices to show that

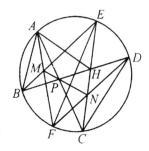

$$\frac{HP}{PB} \cdot \frac{BM}{ME} \cdot \frac{EN}{NH} = 1. \qquad (15.38)$$

Since

$$\frac{BM}{ME} = \frac{[BAF]}{[EAF]} = \frac{BA \cdot BF}{EA \cdot EF}, \qquad (15.39)$$

$$\frac{HP}{PB} = \frac{[HAC]}{[BAC]} = \frac{[HAC]}{[EAC]} \cdot \frac{[EAC]}{[BAC]}$$

$$= \frac{CH}{CE} \cdot \frac{EA \cdot EC}{BA \cdot BC} = \frac{CH \cdot EA}{BA \cdot BC}, \qquad (15.40)$$

therefore

$$\frac{HP}{PB} \cdot \frac{BM}{ME} = \frac{CH \cdot BF}{BC \cdot EF},$$

i.e., it suffices to shoow that

$$\frac{EN \cdot BF \cdot CH}{EF \cdot BC \cdot NH} = 1. \qquad (15.41)$$

Note that $\dfrac{EN}{EF} = \dfrac{DN}{DC}$ and $\angle BFD = \angle BCD$, so

$$\frac{NH}{CH} = \frac{[NBD]}{[CBD]} = \frac{[FBD] - [FBN]}{[CBD]} = \frac{FB \cdot FD - FB \cdot FN}{CB \cdot CD}$$

$$= \frac{FB \cdot ND}{CB \cdot CD} = \frac{FB}{CB} \cdot \frac{EN}{EF},$$

so (15.41) is true, thus, (15.38) is true and the conclusion is proven.

2. Let the six tangent points on circles be $B_1, B_2, B_3, B_4, B_5, B_6$ respectively, as shown in the diagram below.

Suppose that the lines $B_1 B_2, B_3 B_4,$ $B_5 B_6$ intersect pairwise at $P, Q, R.$ Connect $B_1 B_6, B_4 B_5, B_2 B_3.$ By the trigonometric form of Cave's theorem, it follows that

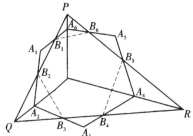

$$\frac{\sin \angle B_1 P A_6}{\sin \angle A_6 P B_6} \cdot \frac{\sin \angle P B_6 A_6}{\sin \angle A_6 B_6 B_1}$$

$$\cdot \frac{\sin \angle B_6 B_1 A_6}{\sin \angle A_6 B_1 P} = 1. \text{ Since } A_6 B_6 = A_6 B_1, \text{ so } \angle A_6 B_6 B_1 = \angle A_6 B_1 B_6, \text{ it}$$

follows that

$$\frac{\sin \angle B_1 P A_6}{\sin \angle A_6 P B_6} \cdot \frac{\angle P B_6 A_6}{\sin \angle A_6 B_1 P} = 1.$$

Similarly,

$$\frac{\sin \angle B_5 R A_4}{\sin \angle A_4 R B_4} \cdot \frac{\angle R B_4 A_4}{\sin \angle A_4 B_5 R} = 1, \quad \frac{\sin \angle B_3 Q A_2}{\sin \angle A_2 Q B_2} \cdot \frac{\angle Q B_2 A_2}{\sin \angle A_2 B_3 Q} = 1.$$

multiplying these three equalities, and considering

$$\angle P B_6 A_6 = \angle A_5 B_6 B_5 = \angle A_5 B_5 B_6 = \angle R B_5 A_4,$$
$$\angle R B_4 A_4 = \angle A_3 B_3 B_3 = \angle A_3 B_3 B_4 = \angle Q B_3 A_2,$$
$$\angle Q B_2 A_2 = \angle A_1 B_2 B_1 = \angle A_1 B_1 B_2 = \angle P B_1 A_6,$$

it follows that

$$\frac{\sin \angle B_1 P A_6}{\sin \angle A_6 P B_6} \cdot \frac{\sin \angle B_5 R A_4}{\sin \angle A_4 R B_4} \cdot \frac{\sin \angle B_3 Q A_2}{\sin \angle A_2 Q B_2} = 1.$$

By the sufficiency of the Ceva's theorem in trigonometric form, the lines PA_6, QA_2, RA_4 are concurrent, namely, the lines e, f, g are concurrent.

3. Let P, Q, R be the midpoints of DF, EF, ED respectively, then $PQ \parallel DE$ and the line PQ passes through N; $RQ \parallel DF$, and the line RQ passes M; $PR \parallel EF$, and the line PR passes through L.

Since $AE = AF$, so AQ is the angle bisecter of $\angle CAB$. Similarly, BP, CR are the angle bisectors of $\angle ABC, \angle BCA$ respectively, and the lines AQ, BP, CR are concurrent at the incenter I of $\triangle ABC$.

Applying the Desargues' Theorem to $\triangle ABC$ and $\triangle QPR$, it follows that the point L (as the point of intersection of PR and BC), the point M (as the point of intersection of RQ and CA), the point N (as the point of intersection of QP and AB) are collinear.

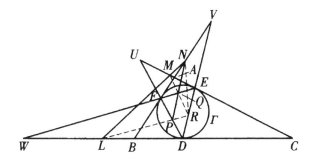

4. As shown in the digram below, from center O introduce $OD \perp AB$ at D and $OE \perp BC$ at E. Let $AB = a, BC = b, OD = d, OE = e$, then $a > b$ and

$$BD = \frac{a}{2}, BE = \frac{b}{2}, a^2 + 4d^2 = 25^2.$$

Similarly, $b^2 + 4e^2 = 25^2$. Since $a, b, c, e \in \mathbb{N}$. If the positive integer solutions $(x, 2y)$ for the equation

$$x^2 + (2y)^2 = 25^2$$

satisfy $(x, 2y) = 1$, then there are positive integers u, v with $u > v$ such that

$$x = u^2 - v^2, y = uv, u^2 + v^2 = 25,$$

so $u = 4, v = 3$, i.e. $x = 7, y = 12$. If $x = 5x_1, y = 5y_1$, then $x_1^2 + (2y_1)^2 = 25$, so $x_1 = 3, 2y_1 = 4$, i.e. $x = 15, y = 10$. Since $a > b$, so $AB = 15, BC = 7, OD = 10, OE = 12$.

Since $OD \perp AB$ and $OE \perp BC$ implies that $ODBE$ is cyclic, By the Ptolemy's theorem,

$$DE \cdot OB = OD \cdot BE + OE \cdot DB \Rightarrow DE = \frac{2}{25}(10 \cdot \frac{7}{2} + 12 \cdot \frac{15}{2}) = 10.$$

Since AC is a midline of $\triangle BAC$, so $AC = 2DE = 20$. Thus, $AB = 15, BC = 7, AC = 20$.

5. Let $\dfrac{BA'}{A'C} = \dfrac{CB'}{B'A} = \dfrac{AC'}{C'B} = k, \dfrac{PQ}{B'C'} = t$. It suffices to show that $t \geq 2$.

$PQ \parallel B'C'$ implies that $\dfrac{AP}{AB'} = \dfrac{AQ}{AC'} = t$. Without loss of generality we may assume that $AP \geq AC, AQ \leq AB$. Then

$$\frac{AQ}{QB} = \frac{AQ}{AB - AQ} = \frac{tAC'}{AB - tAC'} = \frac{t}{\frac{AB}{AC'} - t} = \frac{t}{\frac{1+k}{k} - t}, \quad (15.42)$$

$$\frac{CP}{PA} = \frac{AP - AC}{AP} = \frac{tAB' - AC}{tAB'} = 1 - \frac{AC}{tAB'} = 1 - \frac{1+k}{t}. \quad (15.43)$$

Applying the Menelaus' Theorem to $\triangle ABC$ and the transversal $PA'Q$ gives that $\dfrac{AQ}{QB} \cdot \dfrac{BA'}{A'C} \cdot \dfrac{CP}{PA} = 1$. Substituting (15.42) and (15.43) into it yields

$$\frac{t}{\frac{1+k}{k} - t} \cdot k \cdot \frac{t - (1+k)}{t} = 1 \Rightarrow kt - k(1+k) = \frac{1+k}{k} - t$$

$$\Rightarrow t(1+k) = \frac{1}{k}(1+k) + k(1+k) \Rightarrow t = \frac{1}{k} + k \Rightarrow t \geq 2.$$

6. Let R be the radius of the circle. By the extended sine rule,

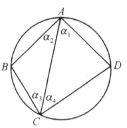

$$AC = 2R\sin(\alpha_4 + \alpha_1) = 2R\sin(\alpha_2 + \alpha_3),$$
$$BD = 2R\sin(\alpha_1 + \alpha_2) = 2R\sin(\alpha_3 + \alpha_4),$$
$$AB = 2R\sin\alpha_3, \quad BC = 2R\sin\alpha_2,$$
$$CD = 2R\sin\alpha_1, \quad DA = 2R\sin\alpha_4.$$

so the given inequality is equivalent to

$$AC^2 \cdot BD^2 \geq 4AB \cdot BC \cdot CD \cdot DA. \qquad (*)$$

The Ptolrmy's Theorem gives

$$AC \cdot BD = AB \cdot CD + BC \cdot AD,$$

so the mean inequality gives

$$AC \cdot BD \geq 2\sqrt{AB \cdot BC \cdot CD \cdot DA},$$

namely $AC^2 \cdot BD^2 \geq 4AB \cdot BC \cdot CD \cdot DA$, $(*)$ is proven.

Testing Questions   (14-B)

1. We prove the conclusion by contradiction. Suppose that A, B, D, C are not
    concyclic, and the circumcircle of $\triangle ABC$ in-
    tersects the line AD at E. Suppose that the
    line BE intersects the line AN at Q, and the
    line CE intersects the line AM at P. Con-
    nect PQ, as shown in the right diagram.
    Below we show that with respect to circle
    $\odot O$ of radius r,

    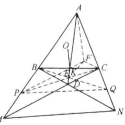

$$PK^2 = \text{the power of } P - \text{the power of } K$$
$$= (PO^2 - r^2) - (r^2 - KO^2).$$

Suppose that the ray PK intersect the circumcircle of $\triangle PAE$ at F. Then

$$PK \cdot KF = AK \cdot KE, \qquad (15.44)$$

and $\angle PFE = \angle PAE = \angle BCE$, so E, C, F, K are concyclic, hence

$$PK \cdot PF = PE \cdot PC. \qquad (15.45)$$

By (15.45) − (15.44), it follows that

$$PK^2 = PE \cdot PC - AK \cdot KE = \text{the power of } P - \text{the power of } K.$$

Similarly, $QK^2 = (QO^2 - r^2) - (r^2 - KO^2)$. Thus, $PO^2 - PK^2 = QO^2 - QK^2$, namely $OK \perp PQ$. Since $OK \perp MN$, so $PQ \parallel MN$, hence

$$\frac{AQ}{QN} = \frac{AP}{PM}. \qquad (15.46)$$

Applying the Menelaus' theorem to $\triangle NDA$ and the transversal BEQ, then

$$\frac{NB}{BD} \cdot \frac{DE}{EA} \cdot \frac{AQ}{QN} = 1. \qquad (15.47)$$

Applying the Menelaus' theorem to $\triangle MDA$ and the transversal CEP, then

$$\frac{MC}{CD} \cdot \frac{DE}{EA} \cdot \frac{AP}{PM} = 1. \qquad (15.48)$$

The combination of (15.46), (15.47), (15.48) then yields $\dfrac{NB}{BD} = \dfrac{MC}{CD}$, so $\dfrac{ND}{BD} = \dfrac{MD}{DC}$ and $\triangle DMN \sim \triangle DCB$, then $\angle DMN = \angle DCB$, i.e., $BC \parallel MN$, so $OK \perp BC$, namely K is the midpoint of BC, a contradiction! Thus, A, B, D, C are concyclic.

2. As shown in the diagram below, let the circumcircles of $\triangle BCE$ and $\triangle CDF$ intersect at C and M. Suppose that P, Q, R be the feet of the perpendiculars from M onto the lines BE, EC and BC respectively, then the Simson's theorem shows that P, Q, R are collinear. Similarly, suppose Q, R, S be the feet of the perpendiculars from M to the lines DC, CF, DF respectively, then Q, R, S are collinear. Therefore P, Q, R, S are collinear.
In the $\triangle ADE$, the collinear three points

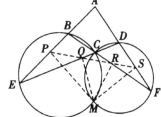

P, Q, S are on its sides AE, DE, AD respectively, so M must be on the circumcircle of $\triangle ADE$ by the sufficiency of Simson's theorem. By the same reasoning it is obtained that M is on the circumcircle of $\triangle ABF$. Thus, the four circumcircles of triangles BCE, CDF, ADE, ABF are concurrent at M.

Note: M is called the *Miquel point*, and the line passing through points P, Q, R, S is called the *Simson's line* of the quadrilateral $ABCD$.

3. Let $\angle BAQ = \alpha, \angle PAN = \beta, \angle PAQ = \gamma$. By applying the Ceva's theorem in trigonometric form to $\triangle ABC$ and P, it follows that

$$\frac{\sin\angle BAP \cdot \sin\angle ACP \cdot \sin\angle CBP}{\sin\angle PAC \cdot \sin\angle PCB \cdot \sin\angle PBA} = 1.$$
(15.49)

Similarly, to $\triangle MNQ$ and P, it follows that

$$\frac{\sin\angle MNP \cdot \sin\angle NQP \cdot \sin\angle QMP}{\sin\angle PNQ \cdot \sin\angle PQM \cdot \sin\angle PMN} = 1.$$
(15.50)

Since $BMPQ$ and $CNPQ$ are both cyclic, so

$$\angle BMQ = \angle BPQ = \angle QCN, \quad \angle QNC = \angle QPC = \angle MBQ.$$

Hence $AMQC$ and $ABQN$ are both cyclic also. Since $MN \parallel BC$, so

$$\angle BAP = \alpha + \gamma, \angle PAC = \beta, \angle PMQ = \angle PBQ = \angle QAN = \beta + \gamma,$$
$$\angle ACP = \angle NCP = \angle NQP, \angle CBP = \angle MNP, \angle PCB = \angle PMN,$$
$$\angle PBA = \angle PQM, \angle PNQ = \angle PCQ = \angle MAQ = \alpha.$$

(15.51) $\div$ (15.52) yields $\dfrac{\sin(\alpha + \gamma)}{\sin\beta} \cdot \dfrac{\sin\alpha}{\sin(\beta + \gamma)} = 1$

$\Leftrightarrow \sin(\alpha + \gamma)\sin\alpha = \sin(\beta + \gamma)\sin\beta$

$\Leftrightarrow \cos\gamma - \cos(2\alpha + \gamma) = \cos\gamma - \cos(2\beta + \gamma)$

$\Leftrightarrow \cos(2\alpha + \gamma) = \cos(2\beta + \gamma) \Leftrightarrow -2\sin(\alpha + \beta + \gamma)\sin(\alpha - \beta) = 0.$

Since $0 < \alpha + \gamma + \beta = \angle BAC < 180°$, so $\alpha = \beta$.

4. The point P is the Miquel's point (cf.Q2 above), so $AEPD$ is cyclic, and

$ABPF$ is cyclic also.

$\angle EAP = \angle EDP = \angle CFP.$ (15.51)

$PEBC$ is cyclic implies that

$$\angle PEA = \angle PCF,$$

so $\triangle AEP \sim \triangle FCP$ and similarly $\triangle EPC \sim \triangle APF$. Therefore

$$\frac{PF}{CP} = \frac{AF}{EC}. \qquad (15.52)$$

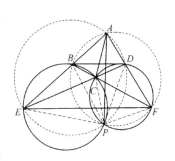

Sufficiency: When $BD \parallel EF$, then $\dfrac{AD}{AF} = \dfrac{BD}{EF} = \dfrac{CD}{EC}$, so from (15.50),

$$\frac{PF}{CP} = \frac{AF}{EC} = \frac{AD}{CD}.$$

Besides, $\angle ADC = \angle FPC$, so $\triangle ACD \sim \triangle FCP \sim \triangle AEP$, hence

$$\angle EAP = \angle CAD, \quad \text{namely} \quad \angle BAP = \angle CAD.$$

Necessity: When $BAP = \angle CAD$, namely $\angle EAP = \angle CAD$, then (15.49) implies $\angle CFP = \angle CAD$. Since $\angle ADC = \angle FPC$, so

$$\triangle CPF \sim \triangle CDA,$$

therefore $\dfrac{PF}{CP} = \dfrac{AD}{CD}$. Then (15.50) implies $\dfrac{AF}{EC} = \dfrac{AD}{CD}$, so $\dfrac{EC}{CD} = \dfrac{AF}{AD}$.
On the other hand, applying the Menelaus' theorem to $\triangle AED$ and the transversal BCF gives

$$\frac{AB}{BE} \cdot \frac{EC}{CD} \cdot \frac{DF}{FA} = 1,$$

so $\dfrac{AB}{BE} \cdot \dfrac{AF}{AD} \cdot \dfrac{DF}{FA} = 1$, i.e., $\dfrac{AB}{BE} = \dfrac{AD}{DF}$, so $BD \parallel EF$.

Solutions To Testing Questions   15

Testing Questions   (15-A)

1. When the line OH passes through a vertex of $\triangle ABC$, the line also passes through the midpoint of opposite side of the vertex, so the conclusion is obvious.

   When the line OH intersects two sides of $\triangle ABC$, say intersects AB and AC, as shown in the right diagram, let G be the center of gravity of $\triangle ABC$. As well known by us, G is on the segment OH, and the line AG passes through the midpoint M of BC. Connect OM, HM. Suppose that $CC' \perp OH$ at C', $MM' \perp OH$ at M' and $BB' \perp OH$ at B'. Then

   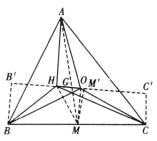

   $$BB' + CC' = 2MM' \Rightarrow [BOH] + [COH] = 2[MOH] = [AOH].$$

   The proofs for other possible locations of the line OH are similar.

2. Connect EF, EC, CD. Then $\angle PEF = \angle PFE = \angle EBF$.
   Since $AF \perp BF$,

   $$\angle BAF = 90° - \angle EBF = 90° - \angle PEF$$
   $$= \tfrac{1}{2}\angle EPF.$$

   Suppose that the circle of center P and radius PE intersects the line BA at A', then

   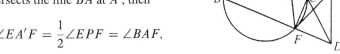

   $$\angle EA'F = \frac{1}{2}\angle EPF = \angle BAF,$$

   so A coincides with A', hence $PA = PE$, and

   $$\angle PAE + \angle ABC = \angle PEA + \angle PEC = 90° \Rightarrow BC \perp AP,$$

   therefore C is the orthocenter of $\triangle ABD$, so $CD \perp AB$. Then $CE \perp AB$ implies that D, C, E are collinear.

3. As shown in the diagram below, $\angle AH_aB = \angle BH_bA = 90°$ implies that

H_a, H_b are on the circle taking F as center. Since R is the midpoint of chord $H_a H_b$, so RF is the perpendicular bisector of $H_a H_b$. Similarly, QE, PD are the perpendicular bisectors of $H_c H_a$, $H_b H_c$ respectively. Thus, PD, QE, RF are concurrent at the circumcenter of $\triangle H_a H_b H_c$.

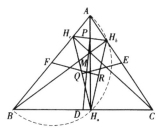

Note: The conclusion is still true when the pedal triangle is not inside the triangle.

4. Let Q be the midpoint of EF. Then $QI \perp EF$. Let H' be the orthocenter of $\triangle DEF$. I is the circumcenter of $\triangle DEF$ then yields

$$IQ = \frac{1}{2} DH'.$$

Since $DP = \frac{1}{2} DM$, By division it follows that

$$\frac{IQ}{DP} = \frac{DH'}{DM}.$$

Since $BH \perp CI$, so

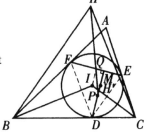

$$\angle HBC = 90° - \angle ICB = 90° - \frac{1}{2}\angle ACB$$
$$= \frac{1}{2}\angle ABC + \frac{1}{2}\angle BAC = \angle IFD + \angle IFE = \angle EFD.$$

Similarly, $\angle HCB = \angle FED$. Hence $\triangle HBC \sim \triangle DFE$. Since I is the orthocenter of $\triangle HBC$, so

$$\frac{DH'}{DM} = \frac{HI}{HD} \Rightarrow \frac{IQ}{DP} = \frac{HI}{HD}.$$

Since $IQ \parallel DP$ and H, I, D are collinear, so $\triangle HIQ \sim HDP$, therefore $\angle IHQ = \angle DHP$, i.e., H, Q, P are collinear, the conclusion is proven.

5. Suppose that the line l_1, l_2, l_3 are such that $l_1 \perp AB$ at M, $l_2 \perp BC$ at N and $l_3 \perp CA$ at P. Below we prove that l_1, l_2, l_3 are concurrent at a point X.

    Suppose that l_1, l_2 intersect at a point X and $XP' \perp AC$ at P'. It suffices to show that P' coincides with P. Let $CP' = x$. Let $p = \frac{1}{2}(a + b + c)$,

then

$$AM = p - b, BN = p - c, CP = p - a, CN = p - b, AP' = b - x,$$
$$BM = p - a. \text{ Then } AM^2 + BN^2 + CP'^2 = CN^2 + BM^2 + AP'^2$$
$$\Rightarrow (p - b)^2 + (p - c)^2 + x^2 = (p - b)^2 + (p - a)^2 + (b - x)^2$$
$$\Rightarrow x = p - a = CP,$$

therefore $P = P'$. Thus, l_3 passes through X, and $AMXP$ is cyclic. Since $AMNP$ is cyclic, so $N = X$, therefore AN is the diameter of the circumcircle of $AMNP$.

(i) Let I_B, I_C be the centers of the escribed circles opposite to B and C respectively, then the application of the Pappus' theorem (cf. p.105) to the two lines $I_C A I_B$ and BNC, it is obtained that M, P, I are collinear.

(ii) Suppose that the incircle $\odot I$ touches AB, AC at R, Q respectively, then the perpendicular bisectors of AB coincides with that of the line segment RM, and they all pass through the midpoint of IN. Similarly, the perpendicular bisector of AC passes through the midpoint of IN also, hence O is the midpoint of IN. Thus, I, O, N are collinear.

6. (i) Connect AH and make $OD \perp BC$ at D. Since O is the circumcenter and $\angle BAC = 60°$,

$$\angle BOC = 2\angle BAC = 120°,$$

$$OD = OC \cos 60° = \frac{1}{2}.$$

The property of Euler line gives $AH = 2OD = 1$. From the extended sine rule

$$BC = 2R \sin \angle BAC = \sqrt{3},$$

so that the area of $ABHC$ is $\frac{1}{2}AH \cdot BC = \frac{\sqrt{3}}{2}$.

(ii) Since H is the orthocenter of $\triangle ABC$,

$$\angle BHC = 180° - (\angle HBC + \angle HCB) = 180° - \angle BAC = 120°.$$

Therefore $BCHO$ is cyclic, $PO \cdot PH = PB \cdot PC$. Since O is the circumcenter of $\triangle ABC$, by considering the power of P to $\odot O$,

$$PB \cdot PC = PO^2 - R^2 \Rightarrow PO \cdot PH = PO^2 - R^2$$
$$\Rightarrow PO^2 - PO \cdot PH = R^2 = 1 \Rightarrow PO \cdot OH = 1.$$

7. Let D, E, F be the midpoints of the sides BC, CA, AB. Since $B_1C_1, C_1A_1,$ A_1B_1 are the perpendicular bisectors of the segments AG, BG, CG respectively, then A_1, B_1, C_1 are the circumcenters of $\triangle GBC, \triangle GCA, \triangle GAB$ respectively. Therefore A_1D, B_1E, C_1F are the perpendicular bisectors of BC, CA, AB respectively, hence the lines A_1D, B_1E, C_1F intersect at O.

It suffices to show that A_1D, B_1E, C_1F are the three medians of $\triangle A_1B_1C_1$.

Suppose that M is the midpoint of AG and the extension of A_1D intersects B_1C_1 at N. We prove that N is the midpoint of B_1C_1 below.

$\angle AMB_1 = \angle AEB_1 = 90°$ implies that $AMEB_1$ is cyclic, therefore $\angle MAE = \angle MB_1E$. Similarly, $CDOE$ is cyclic implies $\angle ECD = \angle EON$. Therefore $\triangle ADC \sim \triangle B_1NO$, so

$$\frac{NB_1}{NO} = \frac{AD}{CD}. \tag{15.53}$$

$\angle AMC_1 = \angle AFC_1 = 90°$ implies that $AMFC_1$ is cyclic, so that $\angle MAF = \angle MC_1F$; $\angle ODB = \angle OFB = 90°$ implies that $DOFB$ is cyclic, so that $\angle FBD = \angle FON$. Therefore $\triangle ADB \sim \triangle C_1NO$, so

$$\frac{NC_1}{NO} = \frac{AD}{BD}. \tag{15.54}$$

Since $CD = BD$, the combination of (15.53) and (15.54) yields $NB_1 = NC_1$, so N is the midpoint of B_1C_1. Similarl, the lines B_1E and C_1F are the other two medians of $\triangle A_1B_1C_1$. Thus, O is the centroid of $\triangle A_1B_1C_1$.

Testing Questions   (15-B)

1. Let K, L, M, B', C' be the midpoints of BP, CQ, PQ, CA, and AB, respectively (see the given diagram below). Since $CA \parallel LM$, we have $\angle LMP = \angle QPA$. Since k touches the line segment PQ at M, we find $\angle LMP = \angle LKM$. Thus $\angle QPA = \angle LKM$. Similarly it follows from $AB \parallel MK$ that $\angle PQA = \angle KMQ = \angle KLM$. Therefore, triangles APQ and MKL are similar, hence

$$\frac{AP}{AQ} = \frac{MK}{ML} = \frac{\frac{1}{2}QB}{\frac{1}{2}PC} = \frac{QB}{PC}. \tag{15.55}$$

Now (15.55) is equivalent to $AP \cdot PC = AQ \cdot QB$ which means that the power of points P and Q with respect to the circumcircle of $\triangle ABC$ are equal, hence $OP = OQ$.

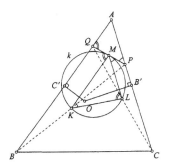

Comment. The last argument can also be established by the following calculation:

$$OP^2 - OQ^2 = OB'^2 + B'P^2 - OC'^2 - C'Q^2$$
$$= (OA^2 - AB'^2) + B'P^2 - (OA^2 - AC'^2) - C'Q^2$$
$$= (AC'^2 - C'Q^2) - (AB'^2 - B'P^2)$$
$$= (AC' - C'Q)(AC' + C'Q) - (AB' - B'P)(AB' + B'P)$$
$$= AQ \cdot QB - AP \cdot PC.$$

With (15.55), we conclude $OP^2 - OQ^2 = 0$, as desired.

2. Connect XY, DX. $BPDX$ and $CYQD$ are both cyclic implies that
$$\angle AXM = \angle BXP = \angle BDP$$
$$= \angle QDC = \angle AYN.$$
$$\because \angle AMX = \angle ANY = 90°,$$
$$\therefore \triangle AMX \sim \triangle ANY,$$

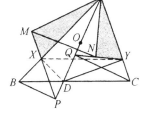

hence $\angle MAX = \angle NAY$ and $\dfrac{AM}{AX} = \dfrac{AN}{AY}$, so $\angle MAN = \angle XAY$ and $\triangle AMN$ is similar to $\triangle AXY$.

Thus, $\triangle AMN \sim \triangle ABC \Leftrightarrow \triangle AXY \sim \triangle ABC \Leftrightarrow XY \parallel BC \Leftrightarrow \angle DXY = \angle XDB$.

$\angle AXD = \angle AYD = 90°$ implies that $AXDY$ is cyclic, so $\angle DXY = \angle DAY$. On the other hand, $\angle XDB = 90° - \angle ABC$, therefore

$$\angle DXY = \angle XDB \Leftrightarrow \angle DAC = 90° - \angle B$$

which is equivalent to that AD passes through the circumcircle of $\triangle ABC$.

3. Let M be the second point of intersection of the circumcircles of $\triangle ACE$ and $\triangle BCD$. It suffices to show that the circumcircle of $\triangle OCI$ also passes through M, since the three circumcenters are all on the perpendicular bisector of CM in that case.

Let $\angle CAB = \alpha, \angle CBA = \beta$. WLOG, we assume that $\beta > \alpha$. Then $\angle OBE = \alpha, \angle DAE = \beta$.

In the quadrilateral $OBIA$, $\angle OAI = \angle OBI = 90°$, so it is cyclic, hence $\angle OIA = \angle OBA = \alpha + \beta$, and

$$\angle CIO = \angle CIA - \angle OIA = 2\angle CBA - (\alpha + \beta) = \beta - \alpha.$$

Since the quadrilateral $AECM$ and $DBMC$ both are cyclic, so that

$$\begin{aligned}\angle BME &= \angle BMC + \angle CME = (180° - \angle CDB) + \angle CAE \\ &= \angle ODA + \angle DAO = 180° - \angle EOB,\end{aligned}$$

therefore the quadrilateral $EOBM$ is cyclic also. Hence

$$\angle CMO = \angle CME - \angle OME = \angle CAE - \alpha = \beta - \alpha = \angle CIO,$$

therefore O, C, M, I are concyclic, the conclusion is proven.

4. (i) Let the orthocenter of $\triangle ABT$ be H and BH intersects AC at K, then $HK \perp TC$ at K. Since point H is circumcenter of $\triangle CDT$, so KH is the perpendicular bisector of TC, therefore $\triangle TBC$ is isosceles with $BT = BC$ and $\angle BCT = \angle BTC$.
Similarly, $\angle ATD = \angle ADT$, therefore

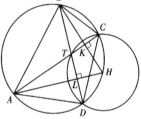

$$\begin{aligned}\angle ADB &= \angle ADT = \angle ATD = \angle BTC \\ &= \angle BCT = \angle BCA,\end{aligned}$$

thus, $ABCD$ is cyclic.

(ii) Suppose that HA intersects BD at L, then $HL \perp TD$ at L, as shown in above diagram. Since $HK \perp TC$ at K, so $LTKH$ is cyclic, therefore

$$\angle AHB = \angle LHK = 180° - \angle LTK = \angle BTC = \angle BCT = \angle BCA.$$

Thus, $ABCH$ is cyclic, namely the circumcenter of $\triangle CDT$ is on the circumcircle of $ABCD$.

5. First of all we prove that $\triangle O_1 O_2 O_3 \sim \triangle ABC$.

As shown in the left diagram below, draw the circumcircles $\odot O_1, \odot O_3$ of triangles AEF, CDE respectively. Besides E, let M be the second point of intersection of $\odot O_1$ and $\odot O_3$. Connect MD, ME, MF. Then

$$\angle B + \angle FMD = \angle B + (\angle FME + \angle EMD) = \angle B + \angle A + \angle C = 180°$$

implies that $BFMD$ is cyclic, i.e., the circumcircle $\odot O_2$ passes through M also.

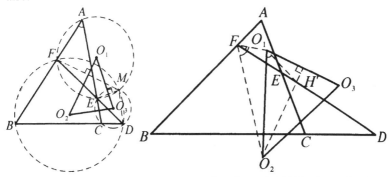

But MF is the radical axis of $\odot O_1$ and $\odot O_2$ and ME is the radical axis of $\odot O_1$ and $\odot O_3$, so $O_1 O_2 \perp MF$ and $O_1 O_3 \perp ME$, therefore $\angle O_1 = \angle FME = \angle A$. Similarly, $\angle O_2 = \angle B$. Thus, $\triangle O_1 O_2 O_3 \sim \triangle ABC$.

Below we return to the original problem: to prove that the orthocenter of $\triangle O_1 O_2 O_3$ is on the line l.

As shown in the right diagram above, from O_2 introduce the perpendicular to $O_1 O_3$, intersecting l at H'. Connect $FO_1, FO_2, O_1 H'$. Since O_1 is the circumcenter of $\triangle AEF$, it's easy to see that $\angle H' FO_1 = \angle O_1 FE = 90° - \angle A$; on the other hand, $\angle H' O_2 O_1 = 90° - \angle O_3 O_1 O_2 = 90° - \angle A$. Thus, $\angle H' FO_1 = \angle H' O_2 O_1$, so $H' O_2 FO_1$ is cyclic, and hence

$$\angle H' O_1 O_2 = \angle H' FO_2.$$

But O_2 is the circumcenter of $\triangle BFD$, so

$$\angle H' FO_2 = \angle O_2 FD = 90° - \angle B.$$

Thus, $\angle H' FO_2 = 90° - \angle B = 90° - \angle O_1 O_2 O_3$, namely

$$\angle H' O_1 O_2 + \angle O_1 O_2 O_3 = 90°,$$

i.e., $O_1 H' \perp O_2 O_3$, so H' is the orthocenter of $\triangle O_1 O_2 O_3$, thus, the conclusion is proven.

Trigonometric Identities

Identities and Their Connections

1. **Fundamental identities:**

$$\sin^2 A + \cos^2 A = 1, \quad 1 + \tan^2 A = \sec^2 A, \quad 1 + \cot^2 A = \csc^2 A,$$
$$\sin(\pi - A) = \sin A, \quad \cos(\pi - A) = -\cos A, \quad \tan(\pi - A) = -\tan A,$$
$$\sin\left(\frac{\pi}{2} - A\right) = \cos A, \quad \cos\left(\frac{\pi}{2} - A\right) = \sin A, \quad \tan\left(\frac{\pi}{2} - A\right) = \cot A.$$
$$\sin(-A) = -\sin A, \quad \cos(-A) = \cos A, \quad \tan(-A) = -\tan A.$$

2. **Addition Formulas:** (*Formulas for Sum of Angles and Difference of Angles*)

   (i) $\sin(A + B) = \sin A \cos B + \cos A \sin B$.

   Proof

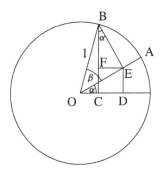

Let $OA = OB = R = 1$, then $\sin(\alpha + \beta) = BC = CF + BF$ and

$$BE = \sin\beta, \quad OE = \cos\beta, \quad ED = OE \sin\alpha = \sin\alpha\cos\beta.$$

Since $\angle FBE = \alpha$,

$$\sin(\alpha + \beta) = CF + BF = ED + BE\cos\alpha = \sin\alpha\cos\beta + \cos\alpha\sin\beta.$$

225

(ii) $\sin(A - B) = \sin A \cos B - \cos A \sin B$.

Proof $\sin(A - B) = \sin[A + (-B)] = \sin A \cos(-B) + \cos A \sin(-B)$
$$= \sin A \cos B - \cos A \sin B.$$

(iii) $\cos(A + B) = \cos A \cos B - \sin A \sin B$.

Proof $\cos(A + B) = \sin[\frac{\pi}{2} - (A + B)] = \sin[(\frac{\pi}{2} - A) + (-B)]$
$$= \sin(\tfrac{\pi}{2} - A)\cos(-B) + \cos(\tfrac{\pi}{2} - A)\sin(-B)$$
$$= \cos A \cos B - \sin A \sin B.$$

(iv) $\cos(A - B) = \cos A \cos B + \sin A \sin B$.

Proof $\cos(A - B) = \cos[A + (-B)] = \cos A \cos(-B) - \sin A \sin(-B)$
$$= \cos A \cos B + \sin A \sin B.$$

(v) $\tan(A + B) = \dfrac{\tan A + \tan B}{1 - \tan A \tan B}$

Proof $\tan(A + B) = \dfrac{\sin A \cos B + \cos A \sin B}{\cos A \cos B - \sin A \sin B}$
$$= \dfrac{(\sin A \cos B + \cos A \sin B)/\cos A \cos B}{(\cos A \cos B - \sin A \sin B)/\cos A \cos B}$$
$$= \dfrac{\tan A + \tan B}{1 - \tan A \tan B}.$$

(vi) $\tan(A - B) = \dfrac{\tan A + \tan(-B)}{1 - \tan A \tan(-B)} = \dfrac{\tan A - \tan B}{1 + \tan A \tan B}.$

3. Formulas for Double Angle, Half Angle and Multiple Angle:

(i) $\sin 2A = 2\sin A \cos A = \dfrac{2\tan A}{1 + \tan^2 A}$

Proof   Letting $B = A$ in (2.(i)) yields $\sin 2A = 2\sin A \cos A$ at once.
Further
$$2\sin A \cos A = 2\frac{\sin A}{\cos A} \cdot \cos^2 A = \frac{2\tan A}{1/\cos^2 A} = \frac{2\tan A}{\sec^2 A} = \frac{2\tan A}{1 + \tan^2 A}.$$

(ii) $\cos 2A = \cos^2 A - \sin^2 A = 2\cos^2 A - 1 = 1 - 2\sin^2 A = \dfrac{1 - \tan^2 A}{1 + \tan^2 A}$

Proof   Letting $B = A$ in (2.(iii)) gives $\cos 2A = \cos^2 A - \sin^2 A$ at once.
Further
$$\cos^2 A - \sin^2 A = \cos^2 A - (1 - \cos^2 A) = 2\cos^2 A - 1 \text{ and}$$
$$\cos^2 A - \sin^2 A = (1 - \sin^2 A) - \sin^2 A = 1 - 2\sin^2 A, \text{ and}$$
$$\cos^2 A - \sin^2 A = \cos^2 A(1 - \tan^2 A) = \frac{1 - \tan^2 A}{\sec^2 A} = \frac{1 - \tan^2 A}{1 + \tan^2 A}.$$

(iii) $\sin^2 A = \dfrac{1}{2}(1 - \cos 2A),$ $\cos^2 A = \dfrac{1}{2}(1 + \cos 2A).$

Proof $\cos 2A = 1 - 2\sin^2 A \Rightarrow 2\sin^2 A = 1 - \cos 2A$
$$\Rightarrow \sin^2 A = \tfrac{1}{2}(1 - \cos 2A);$$
$$\cos 2A = 2\cos^2 A - 1 \Rightarrow 2\cos^2 A = 1 + \cos 2A$$
$$\Rightarrow \cos^2 A = \tfrac{1}{2}(1 + \cos 2A).$$

(iv) $\quad \tan 2A = \dfrac{2\tan A}{1 - \tan^2 A}$

Proof Letting $B = A$ in (2.(v)) yields $\tan 2A = \dfrac{2\tan A}{1 - \tan^2 A}$ at once.

(v) $\quad \sin 3A = 3\sin A - 4\sin^3 A \qquad \cos 3A = 4\cos^3 A - 3\cos A.$

Proof $\sin 3A = \sin(2A + A) = \sin 2A \cos A + \cos 2A \sin A$
$$= 2\sin A \cos^2 A + (1 - 2\sin^2 A)\sin A$$
$$= 2\sin A(1 - \sin^2 A) + \sin A - 2\sin^3 A$$
$$= 3\sin A - 4\sin^3 A;$$

$$\cos 3A = \cos(2A + A) = \cos 2A \cos A - \sin 2A \sin A$$
$$= (2\cos^2 A - 1)\cos A - 2\sin^2 A \cos A$$
$$= 2\cos^3 A - \cos A - 2(1 - \cos^2 A)\cos A = 4\cos^3 A - 3\cos A.$$

(vi) $\quad \tan 3A = \dfrac{3\tan A - \tan^3 A}{1 - 3\tan^2 A}.$

Proof $\tan 3A = \tan(2A + A) = \dfrac{\tan 2A + \tan A}{1 - \tan 2A \tan A}$

$$= \dfrac{\frac{2\tan A}{1-\tan^2 A} + \tan A}{1 - \frac{2\tan^2 A}{1-\tan^2 A}} = \dfrac{3\tan A - \tan^3 A}{1 - 3\tan^2 A}.$$

4. **Formulas for Product to Sum or Difference:**

(i) $\quad 2\sin A \cos B = \sin(A + B) + \sin(A - B)$
Proof The conclusion is obtained at once by $(2.(i)) + (2.(ii))$.

(ii) $\quad 2\cos A \sin B = \sin(A + B) - \sin(A - B)$
Proof The conclusion is obtained at once by $(2.(i)) - (2.(ii))$.

(iii) $\quad 2\cos A \cos B = \cos(A + B) + \cos(A - B)$
Proof The conclusion is obtained at once by $(2.(iii)) + (2.(iv))$.

(iv) $\quad 2\sin A \sin B = \cos(A - B) - \cos(A + B).$
Proof The conclusion is obtained at once by $(2.(iv)) - (2.(iii))$.

5. **Formulas for Sum or Difference to Products:**

(i) $\sin A + \sin B = 2 \sin \dfrac{A + B}{2} \cos \dfrac{A - B}{2}.$

(ii) $\sin A - \sin B = 2 \cos \dfrac{A + B}{2} \sin \dfrac{A - B}{2}.$

(iii) $\cos A + \cos B = 2 \cos \dfrac{A + B}{2} \cos \dfrac{A - B}{2}.$

(iv) $\cos A - \cos B = -2 \sin \dfrac{A + B}{2} \sin \dfrac{A - B}{2}.$

Remark   The formulas in **4** and those in **5** represent same relations actually, but they are applied in opposite direction. For example, if let

$$X = \frac{A + B}{2}, \quad Y = \frac{A - B}{2}$$

in (5.(i)), then $A = X + Y$, $B = X - Y$, so (5.(i)) is equivalent to

$$\sin(X + Y) + \sin(X - Y) = 2 \sin X \cos Y$$

which is the same as the formula (4.(i)). The rest is similar.

6. **R-Formula:**
$$a \cos \theta + b \sin \theta = R \sin(\theta + \alpha),$$
$$a \cos \theta + b \sin \theta = R \cos(\theta + \beta)$$

where
$$R = \sqrt{a^2 + b^2}$$

$$\sin \alpha = \frac{a}{\sqrt{a^2 + b^2}}, \qquad \cos \alpha = \frac{b}{\sqrt{a^2 + b^2}},$$

$$\sin \beta = -\frac{b}{\sqrt{a^2 + b^2}}, \qquad \cos \beta = \frac{a}{\sqrt{a^2 + b^2}}.$$

Proof

$$
\begin{aligned}
a \cos \theta + b \sin \theta &= \sqrt{a^2 + b^2} \left(\frac{a}{\sqrt{a^2 + b^2}} \cos \theta + \frac{b}{\sqrt{a^2 + b^2}} \sin \theta \right) \\
&= R(\sin \alpha \cos \theta + \cos \alpha \sin \theta) = R \sin(\theta + \alpha).
\end{aligned}
$$

$$
\begin{aligned}
a \cos \theta + b \sin \theta &= \sqrt{a^2 + b^2} \left(\frac{a}{\sqrt{a^2 + b^2}} \cos \theta + \frac{b}{\sqrt{a^2 + b^2}} \sin \theta \right) \\
&= R(\cos \beta \cos \theta - \sin \beta \sin \theta) = R \cos(\theta + \beta).
\end{aligned}
$$

Appendix B

Mean Inequality

For any n numbers a_1, a_2, $\cdots$, $a_n > 0$, define the following averages:

Arithmetic Mean   : $A_n = \dfrac{a_1 + a_2 + \cdots + a_n}{n}$,

Geometric Mean   : $G_n = \sqrt[n]{a_1 a_2 \cdots a_n}$,

Harmonic Mean   : $H_n = \dfrac{n}{\dfrac{1}{a_1} + \dfrac{1}{a_2} + \cdots + \dfrac{1}{a_n}}$,

Root-Mean of Squares : $R_n = \sqrt{\dfrac{a_1^2 + a_2^2 + \cdots + a_n^2}{n}}$.

Theorem I. *(Mean Inequality) The sizes of above means have always the following order:*

$$H_n \leq G_n \leq A_n \leq R_n,$$

and any one of the equalities holds if and only if $a_1 = a_2 = \cdots = a_n$.

Proof. $A_n \geq G_n$:   There are a lot of methods for proving it, where many need to use the mathematical induction. For convenience of the readers who have not learned the mathematical induction yet, we introduce the following proof.

Define $f(n) = n \left(\dfrac{a_1 + a_2 + \cdots + a_n}{n} - \sqrt[n]{a_1 a_2 \cdots a_n} \right)$ for $n = 2, 3, 4, \ldots$

Since $A_2 \geq G_2$, it suffices to show that the sequence $\{f(n)\}_{n \geq 2}$ is increasing, i.e.,

$$f(2) \leq f(3) \leq f(4) \leq \cdots \leq f(n) \leq f(n+1) \leq \cdots.$$

Let $a_1 a_2 \cdots a_n = y^{n(n+1)}, a_{n+1} = x^{n+1}$, then $x, y > 0$. From

$$(x - y)(x^k - y^k) \geq 0, \qquad \text{for } k = 1, 2, 3, \ldots$$

229

it follows that for $n \geq 2$

$$x^{n+1} + ny^{n+1} - (n+1)xy^n = x(x^n - y^n) - ny^n(x - y)$$
$$= (x - y)[x(x^{n-1} + x^{n-2}y + \cdots + y^{n-1}) - ny^n]$$
$$= (x - y)[(x^n - y^n) + (x^{n-1} - y^{n-1})y + \cdots + (x - y)y^{n-1}]$$
$$= (x - y)(x^n - y^n) + (x - y)(x^{n-1} - y^{n-1})y + \cdots + (x - y)^2 y^{n-1} \geq 0,$$

The equality holds if and only if $x = y$, i.e. $a_{n+1} = \sqrt[n]{a_1 a_2 \cdots a_n}$. Now

$$
\begin{aligned}
f(n+1) - f(n) &= (n+1)\left(\frac{a_1 + a_2 + \cdots + a_{n+1}}{n+1} - \sqrt[n+1]{a_1 a_2 \cdots a_{n+1}}\right) \\
&\quad -n\left(\frac{a_1 + a_2 + \cdots + a_n}{n} - \sqrt[n]{a_1 a_2 \cdots a_n}\right) \\
&= a_{n+1} - (n+1)\sqrt[n+1]{a_1 a_2 \cdots a_{n+1}} + n\sqrt[n]{a_1 a_2 \cdots a_n} \\
&= x^{n+1} - (n+1)y^n x + ny^{n+1} \geq 0,
\end{aligned}
$$

Thus, $f(n+1) \geq f(n)$ for $n \geq 2$, and $f(n+1) - f(n) = 0$ if and only if $a_{n+1} = \sqrt[n]{a_1 a_2 \cdots a_n}$.

It is obvious that for any positive integer n, $a_1 = a_2 = \cdots = a_n \Rightarrow A_n = G_n$. Conversely, if $A_n = G_n$, then $f(n) = 0$, so $f(2) = f(3) = \cdots \leq f(n) = 0$, which gives that

$$a_{k+1} = \sqrt[k]{a_1 a_2 \cdots a_k}, \qquad \text{for } k = 2, 3, \ldots, n,$$

Since $f(2) = 0$ implies $a_1 = a_2$, so $a_1 = a_2 = \cdots = a_n$ for $n = 2, 3, \ldots$.

$H_n \leq G_n$: For any given $a_1, a_2, \ldots, a_n > 0$ with $n \geq 2$, applying $G_n \leq A_n$ to the n positive numbers $\dfrac{1}{a_1}, \dfrac{1}{a_2}, \cdots, \dfrac{1}{a_n}$ gives

$$\sqrt[n]{\frac{1}{a_1} \cdots \frac{1}{a_n}} \leq \frac{1}{n}\left[\frac{1}{a_1} + \frac{1}{a_2} + \cdots + \frac{1}{a_n}\right],$$

i.e., $\dfrac{1}{G_n} \leq \dfrac{1}{H_n}$. Therefore $H_n \leq G_n$.

The condition for holding the equality then is obvious from above proof of $G_n \leq A_n$.

$A_n \leq R_n$: $\dfrac{a_1 + a_2 + \cdots + a_n}{n} \leq \sqrt{\dfrac{a_1^2 + a_2^2 + \cdots + a_n^2}{n}}$

$\Leftrightarrow n(a_1^2 + a_2^2 + \cdots + a_n^2) - (a_1 + a_2 + \cdots + a_n)^2 \geq 0$

$\Leftrightarrow (a_1 - a_2)^2 + \cdots + (a_1 - a_n)^2 + (a_2 - a_3)^2 + \cdots + (a_2 - a_n)^2 + \cdots$
$\qquad + (a_{n-2} - a_{n-1})^2 + (a_{n-2} - a_n)^2 + (a_{n-1} - a_n)^2 \geq 0,$

and the last inequality is obvious. The condition for holding the equality is clear also. □

Appendix C

Some Basic Inequalities Involving a Triangle

1. Let ABC be a triangle. Then

(a) $\sin \dfrac{A}{2} \sin \dfrac{B}{2} \sin \dfrac{C}{2} \le \dfrac{1}{8}$;

(b) $\sin^2 \dfrac{A}{2} + \sin^2 \dfrac{B}{2} + \sin^2 \dfrac{C}{2} \ge \dfrac{3}{4}$;

(c) $\cos^2 \dfrac{A}{2} + \cos^2 \dfrac{B}{2} + \cos^2 \dfrac{C}{2} \le \dfrac{9}{4}$;

(d) $\cos \dfrac{A}{2} \cos \dfrac{B}{2} \cos \dfrac{C}{2} \le \dfrac{3\sqrt{3}}{8}$;

(e) $\csc \dfrac{A}{2} + \csc \dfrac{B}{2} + \csc \dfrac{C}{2} \ge 6$.

Proof Considering $0 \le |\frac{B-C}{2}| < 90°$ and the extended sine rule,

$$\frac{a}{b+c} = \frac{\sin A}{\sin B + \sin C} = \frac{2 \sin \frac{A}{2} \cos \frac{A}{2}}{2 \sin \frac{B+C}{2} \cos \frac{B-C}{2}} = \frac{\sin \frac{A}{2}}{\cos \frac{B-C}{2}} \ge \sin \frac{A}{2}.$$

Similarly, it is true that

$$\frac{b}{c+a} \ge \sin \frac{B}{2}, \qquad \frac{c}{a+b} \ge \sin \frac{C}{2},$$

therefore

$$\sin \frac{A}{2} \sin \frac{B}{2} \sin \frac{C}{2} \le \frac{abc}{(a+b)(b+c)(c+a)}.$$

The AM-GM inequality yields

$$(a+b)(b+c)(c+a) \ge (2\sqrt{ab})(2\sqrt{bc})(2\sqrt{ca}) = 8abc.$$

Combining the last two equalities gives part (a).

231

By using the identity $\cos A + \cos B + \cos C = 1 + 4 \sin \dfrac{A}{2} \sin \dfrac{B}{2} \sin \dfrac{C}{2}$ and (a),

$$\sin^2 \frac{A}{2} + \sin^2 \frac{B}{2} + \sin^2 \frac{C}{2} = \frac{3 - \cos A - \cos B - \cos C}{2}$$

$$= 1 - 2 \sin \frac{A}{2} \sin \frac{B}{2} \sin \frac{C}{2} \geq 1 - \frac{1}{4} = \frac{3}{4},$$

so (b) is proven.

The Part (c) is obtained at once from (b) and

$$\cos^2 \frac{A}{2} + \cos^2 \frac{B}{2} + \cos^2 \frac{C}{2} = 3 - \left(\sin^2 \frac{A}{2} + \sin^2 \frac{B}{2} + \sin^2 \frac{C}{2} \right).$$

Finally, by (c) and by the AM-GM inequality, we have

$$\frac{9}{4} \geq \cos^2 \frac{A}{2} + \cos^2 \frac{B}{2} + \cos^2 \frac{C}{2} \geq 3 \sqrt[3]{\cos^2 \frac{A}{2} \cos^2 \frac{B}{2} \cos^2 \frac{C}{2}},$$

implying (d).

Again as beginning of the proof,

$$\csc \frac{A}{2} \geq \frac{b+c}{a} = \frac{b}{a} + \frac{c}{a}, \quad \csc \frac{B}{2} \geq \frac{c}{b} + \frac{a}{b}, \quad \csc \frac{C}{2} \geq \frac{a}{c} + \frac{b}{c}.$$

Then

$$\csc \frac{A}{2} + \csc \frac{B}{2} + \csc \frac{C}{2} \geq \left(\frac{b}{a} + \frac{a}{b} \right) + \left(\frac{c}{b} + \frac{b}{c} \right) + \left(\frac{a}{c} + \frac{c}{a} \right) \geq 6,$$

so the Part (e) is proven.

2.   For any acute triangle ABC, $\tan A \tan B \tan C \geq 3\sqrt{3}$.

Proof   Since $\tan A + \tan B + \tan C = \tan A \tan B \tan C$, by the AM-GM inequality,

$$\tan A \tan B \tan C = \tan A + \tan B + \tan C \geq 3 \sqrt[3]{\tan A \tan B \tan C}$$

which yields $(\tan A \tan B \tan C)^{\frac{2}{3}} \geq 3$, so $\tan A \tan B \tan C \geq 3^{\frac{3}{2}} = 3\sqrt{3}$.

3. For any triangle ABC,

(a) $\cos A + \cos B + \cos C \leq \dfrac{3}{2}$;

(b) $\cos A \cos B \cos C \leq \dfrac{1}{8}$;

(c) $\sin A \sin B \sin C \leq \dfrac{3\sqrt{3}}{8}$;

(d) $\sin A + \sin B + \sin C \leq \dfrac{3\sqrt{3}}{2}$;

(e) $\cos^2 A + \cos^2 B + \cos^2 C \geq \dfrac{3}{4}$;

(f) $\sin^2 A + \sin^2 B + \sin^2 C \leq \dfrac{9}{4}$;

(g) $\cos 2A + \cos 2B + \cos 2C \geq -\dfrac{3}{2}$;

(h) $\sin 2A + \sin 2B + \sin 2C \leq \dfrac{3\sqrt{3}}{2}$.

Solution For part (a), from the double angle formulae and 1(b),

$$\cos A + \cos B + \cos C = 3 - 2\left(\sin^2 \frac{A}{2} + \sin^2 \frac{B}{2} + \sin^2 \frac{C}{2}\right) \leq 3 - \frac{3}{2} = \frac{3}{2},$$

as desired.

For part (b), if triangle ABC is non-acute, the left-hand side of the inequality is non-positive, and so the inequality is clearly true.

If ABC is acute, then $\cos A, \cos B, \cos C$ are all positive. Below we show that $(b) \Leftrightarrow (e)$. For this, first note that

$$\cos 2A + \cos 2B + \cos 2C = 2\cos(A + B)\cos(A - B) + (2\cos^2 C - 1)$$
$$= -1 - 2\cos C \cos(A - B) + 2\cos^2 C$$
$$= -1 - 2\cos C(\cos(A - B) + \cos(A + B))$$
$$= -1 - 4\cos A \cos B \cos C.$$

Therefore $2(\cos^2 A + \cos^2 B + \cos^2 C) = 2 - 4\cos A \cos B \cos C$, i.e.,

$$\cos^2 A + \cos^2 B + \cos^2 C = 1 - 2\cos A \cos B \cos C.$$

Thus, $(b) \Leftrightarrow (e)$ is obtained at once. Then from the AM-GM inequality and (a),

$$\cos A \cos B \cos C \leq \left(\frac{\cos A + \cos B + \cos C}{3}\right)^3 \leq \left(\frac{1}{2}\right)^3 = \frac{1}{8}.$$

Thus, (b), and (e) also, is proven.

The two inequalities in parts (e) and (f) are equivalent because $\cos^2 x + \sin^2 x = 1$.

Then (f) and the AM-GM inequality gives

$$\frac{9}{4} \geq \sin^2 A + \sin^2 B + \sin^2 C \geq 3 \sqrt[3]{\sin^2 A \sin^2 B \sin^2 C},$$

from which (c) follows.

Since $(a-b)^2 + (b-c)^2 + (c-a)^2 \geq 0 \Rightarrow 3(a^2+b^2+c^2) \geq (a+b+c)^2$, from (f),

$$\frac{27}{4} \geq 3(\sin^2 A + \sin^2 B + \sin^2 C) \geq (\sin A + \sin B + \sin C)^2$$

gives (d).

Part (g) follows from (e) since $\cos 2x = 2\cos^2 x - 1$. Finally, (h) follows from (c) and the identity

$$\sin 2A + \sin 2B + \sin 2C = 2\sin(A + B)\cos(A - B) + 2\sin C \cos C$$
$$= 2\sin C(\cos(A - B) - \cos(A + B)) = 4\sin A \sin B \sin C.$$

4.   For any triangle ABC and real numbers x, y, z,

$$x^2 + y^2 + z^2 \geq 2xy \cos a + 2yz \cos B + 2xz \cos C.$$

Proof   Let $f(x) = x^2 - (2y \cos A + 2z \cos C)x + y^2 + z^2 - 2yz \cos B$, where y, z are considered as two parameters, then the curve of $f(x)$ is open upwards. It suffices to show that its discriminant Δ is always non-positive. Since

$$
\begin{aligned}
\Delta &= 4(y \cos A + z \cos C)^2 - 4(y^2 + z^2 - 2yz \cos B) \\
&= -4[y^2 \sin^2 A + z^2 \sin^2 C - 2yz(\cos A \cos C - \cos(A + C))] \\
&= -(y^2 \sin^2 A + z^2 sin^2 C + 2yz \sin A \sin C) \\
&= -(y \sin A + z \sin C)^2 \leq 0,
\end{aligned}
$$

so $f(x) \geq 0$ always.

Appendix D

Proofs of Some Important Theorems in Geometry

Theorem I. *(Power of a Point Theorem) Let $\odot O$ be a circle of radius R. For an interior point P of the circle, if AB is a chord passing through P, then the value of product $PA \cdot PB$ is independent of the choice of the chord. In particular, the constant value is given by taking $AB \perp OP$, so it is*

$$R^2 - OP^2.$$

Proof. Let CD be another chord passing through P. Connect AC, BD. For $\triangle PAC$ and $\triangle PDB$,

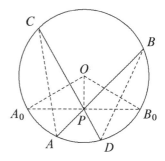

$$\angle PAC = \angle PDB, \quad \angle PCA = \angle PBD,$$

$$\therefore \triangle PAC \sim \triangle PDB,$$

$$\therefore \frac{PA}{PD} = \frac{PC}{PB}$$

i.e. $PA \cdot PB = PC \cdot PD$. In particular, when $A_0 B_0 \perp OP$ at P, then $PA_0 = PB_0$ and

$$PA_0 \cdot PB_0 = (PA_0)^2 = R^2 - OP^2, \quad \text{where } R \text{ is the radius of the circle.} \quad \square$$

Theorem II. *Let $\odot O$ be a circle of radius R. For a point P which is outside the circle or on the circumference, if PAB is the transversal line starting from P and intersecting the circle at A and B, then the value of product $PA \cdot PB$ is independent of the choice of the PAB. In particular, the constant value is given by $(PT)^2$, where the line segment PT is tangent to the circle at T, so it is*

$$OP^2 - R^2.$$

235

Consequence: When line segments AB and CD or their extensions intersect at P, then A, B, C, D are concyclic if and only if $AP \cdot PB = CP \cdot PD$.

Proof.   Connect TA, TB. For $\triangle PAT$ and $\triangle PTB$,

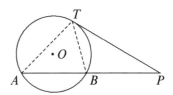

$\angle PAT = \angle PCB$, and the $\angle P$ is shared,

$\therefore \triangle PAT \sim \triangle PTB$,

$$\therefore \frac{PA}{PT} = \frac{PT}{PB}$$

i.e. $PA \cdot PB = PT^2$.

When P is on $\odot O$, then $T = P = B$, so $PA \cdot PB = 0 = (PT)^2$ holds. $\square$

If PDC is another transversal line cutting the circle $\odot O$ at D, C respectively, then, by the result of Theorem II, $PA \cdot PB = (PT)^2 = PC \cdot PD$.

Conversely, When the four points A, B, C, D satisfy $PA \cdot PB = PC \cdot PD$ (where $PC > PD$), then D must on the circumcircle of $\triangle CAB$ ω. Otherwise, let D' be the point of intersection of ω and the line PC, then $PA \cdot PB = PC \cdot PD'$ implies that $PC \cdot PD = PC \cdot PD'$, so $D = D'$, a contradiction.

Thus, the consequence is proven.

Existence of Radical Axis:

(i)   When two circles ω_1 and ω_2 with radii R and r respectively are non-concentric and not intersected, as shown in the following figure:

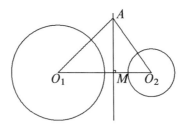

Let the system of coordinates of the points O_1, O_2 be $(0, 0)$ and $(d, 0)$, respectively. When A with coordinates (x, y) has two equal powers to the circles ω_1 and ω_2, then

$$x^2 + y^2 - R^2 = (x - d)^2 + y^2 - r^2,$$
$$x^2 + y^2 - R^2 = x^2 - 2dx + d^2 + y^2 - r^2,$$

$$x = \frac{d^2 + R^2 - r^2}{2d}.$$

Since $d > R + r$, $x - R = \dfrac{(d - R)^2 - r^2}{2d} > 0$, therefore $R < x$. From $0 < x^2 - R^2 = (x - d)^2 - r^2$, we have $x < d - r$, therefore

$$R < x < d - r.$$

We have found that the locus of the variable point A is a straight line perpendicular to the line $O_1 O_2$. If M is the point of intersection of the axis with the line $O_1 O_2$, then $x_M = \dfrac{d^2 + R^2 - r^2}{2d}$.

Remark When $d > 0$, $d + r < R$, then we have similar result as above.

(ii) When the circles ω_1 and ω_2 are tangent externally, then $d = R + r$, so $x_M = R$, i.e. the radical axis is the internal common tangent line of the circles.

(iii) When ω_1 and ω_2 intersect at two points A and B, then A and B are both on the radical axis, so the line AB is the radical axis.

Theorem III. (Menelaus' Theorem) *If a straight line cuts the sides AB, BC and CA (or their extensions) of a $\triangle ABC$ at points X, Y and Z respectively, then*

$$\frac{AX}{XB} \cdot \frac{BY}{YC} \cdot \frac{CZ}{ZA} = 1. \tag{D.1}$$

Proof. For the case shown in the left diagram below, introduce a straight line l such that $l \perp XYZ$ at O. Let the projections of A, B and C on l be A', B', C' respectively. Then

$$\frac{AX}{XB} = \frac{A'O}{OB'}, \qquad \frac{BY}{YC} = \frac{B'O}{OC'}, \qquad \frac{CZ}{ZA} = \frac{C'O}{OA'},$$

therefore the conclusion is obtain by multiplying the three equalities.

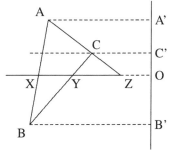
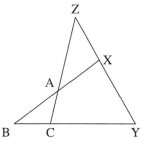

For the case represented by the right diagram above, the proof is similar. □

Theorem IV. *(Inverse Menelaus' Theorem)   For any given $\triangle ABC$, if X, Y, Z are points on lines AB, BC, CA respectively (where exact one point is on the extension of a side, or three points are all on the extensions of sides) such that (D.1) holds, then X, Y, Z must be collinear.*

Proof.   When two points, without loss of generality we may assume that they are X and Y, are on the line segments AB and BC respectively and Z is on the extension of AC, suppose that the line YZ intersects the line segment AB at X'. By the Menelaus' Theorem,

$$\frac{AX'}{X'B} \cdot \frac{BY}{YC} \cdot \frac{CZ}{ZA} = 1,$$

so (D.1) gives $\dfrac{AX}{XB} = \dfrac{AX'}{X'B}$. Then

$$\frac{AX}{XB} = \frac{AX'}{X'B} \Rightarrow \frac{AB}{XB} = \frac{AB}{X'B} \Rightarrow XB = X'B \Rightarrow X \text{ coincides with } X'$$

since X, X' are both on the line segment AB.

For the case that X, Y, Z are all on the extensions of AB, BC, CA respectively, the proof is similar.   □

Theorem V. *(Ceva's Theorem)   For any given triangle ABC, let X, Y, Z be points with (i) all on the line segments BC, CA, AB; or (ii) exact one on one side and other two on the extensions of the two sides respectively. Then the lines AX, BY, CZ are parallel or concurrent if and only if*

$$\frac{BX}{XC} \cdot \frac{CY}{YA} \cdot \frac{AZ}{ZB} = 1. \tag{D.2}$$

Proof.   *Necessity*:   If $AX \parallel BY \parallel CZ$, as shown in left diagram below, then

$$\frac{YC}{YA} = \frac{BC}{BX}, \frac{ZA}{ZB} = \frac{CX}{CB} \Rightarrow \frac{BX}{XC} \cdot \frac{CY}{YA} \cdot \frac{AZ}{ZB} = \frac{BX}{XC} \cdot \frac{BC}{BX} \cdot \frac{CX}{CB} = 1.$$

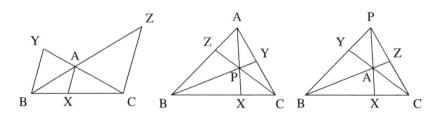

If AX, BY, CZ intersect at a common point P, as shown in the central or right diagrams above, since BPY (or BYP) is a transversal to $\triangle AXC$, by the Menelaus' theorem,

$$\frac{XB}{BC} \cdot \frac{CY}{YA} \cdot \frac{AP}{PX} = 1.$$

Since CPZ (or CZP) is a transversal to $\triangle ABX$,

$$\frac{BC}{CX} \cdot \frac{XP}{PA} \cdot \frac{AZ}{ZB} = 1.$$

Then multiplying the two equalities gives (D.2) at once.

Sufficiency: Now suppose that (D.2) holds.

When the three points X, Y, Z are all on the line segments BC, CA, AB respectively, we can assume that the lines BY and CZ intersects at a point P, so that the line AP intersects BC at some point X'. Then the proof of necessity gives

$$\frac{BX'}{X'C} \cdot \frac{CY}{YA} \cdot \frac{AZ}{ZB} = 1. \tag{D.3}$$

Combination of (D.2) and (D.3) yields $\dfrac{BX'}{X'C} = \dfrac{BX}{XC}$, so $X'C = XC$ which implies $X = X'$ since X', X are both on BC.

When X, Y, Z are not all on the sides of $\triangle ABC$, we may assume that Y and Z are on the extensions of BA and CY respectively, and X is on BC. If $BY \parallel CZ$, by introducing X' on BC such that $AX' \parallel BY$, then it's easy to find that $X' = X$ similarly. If BY and CZ intersect at a point P, then same reasoning as above shows $X' = X$ also. Thus, sufficiency is proven. $\quad\square$

Trigonometric Form of Ceva's Theorem   The Condition (D.2) can be restated as

$$\frac{\sin \angle BAX \cdot \sin \angle CBY \cdot \sin \angle ACZ}{\sin \angle CAX \cdot \sin \angle ABY \cdot \sin \angle BCZ} = 1.$$

Lemma   For each of the following two cases, the following equality holds:

$$\frac{BX}{XC} = \frac{AB}{AC} \cdot \frac{\sin \angle BAX}{\sin \angle CAX}.$$

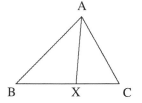 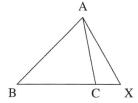

Proof of Lemma.   Applying the sine rule to $\triangle ABX$ and $\triangle ACX$ gives

$$\frac{BX}{AB} = \frac{\sin \angle BAX}{\sin \angle BXA}, \quad \frac{XC}{AC} = \frac{\sin \angle CAX}{\sin \angle CXA},$$

Since $\sin \angle BXA = \sin \angle CXA$, the conclusion is obtained at once by division of these equalities.

From the lemma similarly we have

$$\frac{CY}{YA} = \frac{BC}{AB} \cdot \frac{\sin \angle CBY}{\sin \angle ABY}, \quad \frac{AZ}{ZB} = \frac{AC}{BC} \cdot \frac{\sin \angle ACZ}{\sin \angle BCZ},$$

then multiplying the three equalities yields the conclusion at once.

Theorem VI. (Simson's Theorem)   *For a $\triangle ABC$ and a point D which is outside the triangle, introduce three perpendicular lines from D to the sides BC, CA, intersecting them at A_1, B_1, C_1 respectively. Then A_1, B_1 and C_1 are collinear if and only if A, B, C, D are concyclic.*

Note: When A_1, B_1, C_1 are collinear, the line passing through them is called the **Simson line**.

Proof.   *Sufficiency*:   When $ABCD$ is cyclic, connect DA, DB, DC. Since $C_1 AB_1 D$ and $B_1 A_1 CD$ both are cyclic,

$$\angle CDA_1 = \angle CB_1 A_1 \quad \text{and} \quad \angle C_1 DA = \angle C_1 B_1 A.$$

On the other hand, since $ABCD$ are cyclic,

$$\angle DAC_1 = \angle DCA_1 \Rightarrow \angle C_1 DA = \angle CDA_1$$
$$\Rightarrow \angle C_1 B_1 A = \angle CB_1 A_1$$
$$\Rightarrow C_1, B_1, A_1 \text{ are collinear.}$$

The sufficiency is proven.

Necessity: Suppose that A_1, B_1, C_1 are collinear. Then $\angle C_1 B_1 A = \angle CB_1 A_1$. Since $C_1 AB_1 D$ and $B_1 A_1 CD$ both are cyclic,

$$\angle C_1 DA = \angle CDA_1 \Rightarrow \angle DAC_1 = \angle DCA_1,$$

therefore $ABCD$ is a cyclic quadrilateral. The necessity is proven.   □

Theorem VII. (Ptolemy's Theorem)   *Let $ABCD$ be a convex quadrilateral. Then*

$$AB \cdot CD + AD \cdot BC = AC \cdot BD \tag{D.4}$$

if and only if A, B, C, D are concyclic.

Extended Ptolemy's theorem   *For any convex quadrilateral $ABCD$, the inequality*

$$AB \cdot CD + AD \cdot BC \geq AC \cdot BD \qquad (D.5)$$

always holds, and the equality holds if and only if A, B, C, D are concyclic.

It suffices to show the extended version.

Proof.   *Sufficiency*:   Let the quadrilateral $ABCD$ be cyclic which is inscribed in a circle with radius R. Below we prove that

$$AB \cdot CD + BC \cdot AD = AC \cdot BD.$$

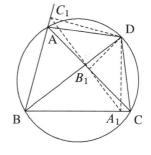

From D introduce perpendiculars to the sides BC, CA, AB of $\triangle ABC$. Let A_1, B_1, C_1 be the perpendicular feet respectively.   Since A, C_1, D, B_1 are concyclic and AD is the diameter of its circumcircle, by the sine rule,

$$B_1 C_1 = AD \sin \angle C_1 A B_1,$$

$$\therefore B_1 C_1 = AD \sin \angle BAC = \frac{AD \cdot BC}{2R}.$$

Similarly, $C_1 A_1 = \dfrac{BD \cdot CA}{2R}$ and $A_1 B_1 = \dfrac{CD \cdot AB}{2R}$. By Simson's theorem, A_1, B_1, C_1 are collinear, i.e., $A_1 B_1 + B_1 C_1 = A_1 C_1$, so

$$CD \cdot \frac{AB}{2R} + AD \cdot \frac{BC}{2R} = BD \cdot \frac{CA}{2R},$$

thus, $CD \cdot AB + AD \cdot BC = BD \cdot CA$.

Necessity: When $CD \cdot AB + AD \cdot BC = BD \cdot CA$, then, by conversing above reasoning, $A_1 B_1 + B_1 C_1 = A_1 C_1$, i.e., A_1, B_1, C_1 are collinear, so D must be on the circumcircle of the $\triangle ABC$ by the Simson's theorem again.

When A, B, C and D are not concyclic, then Simson's theorem shows that A_1, B_1 and C_1 cannot be collinear, therefore $A_1 B_1 + B_1 C_1 > A_1 C_1$, i.e,

$$CD \cdot \frac{AB}{2R} + AD \cdot \frac{BC}{2R} > BD \cdot \frac{CA}{2R},$$

therefore $CD \cdot AB + AD \cdot BC > BD \cdot CA$. $\square$

Theorem VIII. *The three medians of a triangle intersect at one common point, denoted by G as usual, and each median is partitioned by G as two parts of ratio*

2 : 1. *The common point G is called the* **center of gravity** *or* **centroid** *of the triangle.*

Proof. Let G be the point of intersection of the medians AD and BE. From D introduce $DH \parallel BE$, such that DH intersects AC at H. Then

$$CH = HE = \frac{1}{2}AE,$$

$$\therefore AG = 2GD \text{ and } GE = \frac{2}{3}DH = \frac{1}{3}BE,$$

$$\therefore BG = 2GE \text{ also.}$$

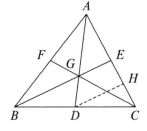

Suppose that CF intersects AD at G', then similarly

$$AG' = 2G'D = \frac{2}{3}AD = AG,$$

hence $G = G'$. Thus AD, BE, CF are concurrent at G. $\square$

Consequence   An interior point P of $\triangle ABC$ is the center of gravity of $\triangle ABC$ if and only if

$$[PBC] = [PCA] = [PAB].$$

Proof By passing through P introduce the lines $l_{bc} \parallel BC$ and $l_{ca} \parallel CA$, then the center of gravity G must be on each of l_{bc} and l_{ca}, so the point of intersection of l_{bc} and l_{ca} must be G, namely $P = G$.

Theorem IX. *For any triangle, the perpendicular bisectors of three sides intersect at a common point O. The O is the center of circumcircle of the triangle, called* **circumcenter** *of the triangle.*

Proof.   For any $\triangle ABC$, if the point of intersection of the perpendicular bisectors of BC and CA is O, then $OB = OC = OA$ implies O is on the perpendicular bisector of the side AB also. $\square$

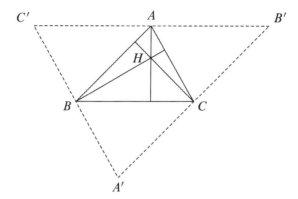

Theorem X. *The three altitudes of any $\triangle ABC$ intersect at one common point H, called* **orthocenter** *of the triangle.*

Proof. As shown in above graph, by passing through each of A, B, C introduce a line parallel to the opposite side, and let $\triangle A'B'C'$ be the triangle formed by these three lines, then the altitudes of $\triangle ABC$ becomes three perpendicular bisectors of sides of $\triangle A'B'C'$, so they intersect at one common point H (i.e. the circumcenter of $\triangle A'B'C'$). □

Theorem XI. *For any triangle, its angle bisectors of three interior angles intersect at one common point, denoted by I as usual, called* **incenter** *(or* **inner center***) of the triangle. I is the center of inscribed circle of the triangle.*

Proof. The point of intersection of any two inner angle bisectors must be on the angle bisector of the third inner angle. □

Theorem XII. *For a triangle, the angle bisectors of one interior angle and two exterior angles of the other two interior angles intersect at common point, called* **excenter** *of the triangle. There are three such points for a triangle, and each is thee center of an escribed circle of the triangle.*

Proof. Similar to Theorem XI. □

Index